# PLASTIC
# WORDS

Uwe Poerksen

# PLASTIC WORDS

## The Tyranny of a Modular Language

Translated by
Jutta Mason and David Cayley

The Pennsylvania State University Press
University Park, Pennsylvania

Library of Congress Cataloging-in-Publication Data

Pörksen, Uwe.
  [Plastikwörter. English]
  Plastic words : the tyranny of a modular language / Uwe Poerksen ;
  translated by Jutta Mason and David Cayley.
    p.     cm.
  Includes bibliographical references (p.      ) and index.
  ISBN 0-271-01476-8
  1. Language and languages—Foreign words and phrases.
  2. Semantics.   3. Science—Terminology.   I. Title.
  P324.P613   1988
  412—dc20                                              94-44589
                                                             CIP

First published in Germany as *Plastikwörter: Die Sprache einer internationalen Diktatur.* © 1988 Klett-Cotta

Published by The Pennsylvania State University Press,
University Park, PA 16802-1003

It is the policy of The Pennsylvania State University Press to use acid-free paper for the first printing of all clothbound books. Publications on uncoated stock satisfy the minimum requirements of American National Standard for Information Sciences—Permanence of Paper for Printed Library Materials, ANSI Z39.48-1984.

*For Ivan and Julian*

The development of Lego went from building blocks that couldn't be attached to one another to building blocks that stuck together, from the static building block to flexible models.

One of the decisive reasons for the worldwide success of Lego is the fact that from 1965 on it was planned as a system:
—all bricks fit together
—kits are compatible
—materials for all age groups are available, starting with "Duplo" for the little ones all the way to "Technic" for the older ones.

(Information plaque for a Lego city in the Deutsches Museum, Munich)

# Contents

# Preface to the
# English Translation

## Translated by Judith Van Herik

Three years ago, as I lectured from *Plastic Words* at Penn State University, I was surprised by my listeners' response. I had chosen my, by then, only dimly remembered Chapter 3, which was also the most "German" of the chapters. Yet, as I read from an early version of this translation, the room filled with laughter. Obviously, the story it told was a familiar one to my listeners, and the translation sounded natural to them. My host, the director of the Science, Technology, and Society Program, ended my presentation by reading excerpts from the University's official internal newspaper to deepen the audience's impression that the language I had described was hardly peculiar to Germany.

As I was writing the book in German, I settled on the title *Lego-words: The Language of a Quiet [leise] Dictatorship [Diktatur]*, but my publisher said that this would not do. First, it evoked the image of Lego blocks in the reader's memory, which was one point of the book, but I was told that the name "Lego" was protected by law and therefore not allowable in a title. I came upon the fairly strong word *Diktatur* by chance. (In German, like in the Latin, *Diktatur* means dictator, dictation, and dictation-giver.) I had lectured on this topic for the first time in a German academy in 1987. In my lecture I analyzed the self-perpetuating dynamics of terms such as "development," "sexuality," "problem," "solution," and "strategy." A woman in the audience asked if I would say that this sort of terminology, used in a particular way, was in practice a dictatorship. I hesitated and then answered: "Yes, a *leise Diktatur*—a quiet, gentle dictatorship." A journalist in the audience incorporated this reply into a short report that was provided to the German Press

Agency: "Professor Speaks on the Language of a Dictatorship in West Germany." The next morning, I received calls from twenty or more newspapers and radio stations, who queried what I had meant and requested interviews. I declined the interviews because it was still too early for me to speak.

At the same time, though, the smart Christian politician Erhard Eppler wrote to request the manuscript. I sent it to him, and soon thereafter sent him the book. He answered that he was leading a heroic and not entirely hopeless battle against plastic language in the new platform of the Social Democratic Party and sadly declared: "Our political language is empty and barely moves anything anymore" ("Unsere politische Sprache ist leer und bewegt fast nichts mehr"). In turn, I chose his words for the title of a lecture that I delivered in 1989, in which I recalled the forgotten German writer Carl Gustav Jochmann, whose 1828 analysis of German political language provided us with insights similar to those in Tocqueville's 1835 essay on how American democracy has changed the English language. So it went, back and forth, between us. This lecture, together with the book *Plastikwoerter*, in fact gave rise to the loveliest experience that I have yet had as a result of something I have written. Erhard Eppler wrote a book that suddenly, in edition after edition, moved "our" topic into widespread and public political discussion: *Kavalleriepferde beim Hornsignal: Die Krise der Politik im Spiegel der Sprache* (Cavalry-horses answering the bugle: The crisis in politics as mirrored in language, 1992).

No one is the first. Already in 1946, George Orwell detected plastic words in his essay "Politics and the English Language." Orwell said then that political speech contained ever fewer words that were carefully chosen for their meaning but ever more phrases tacked together like the sections of a prefabricated henhouse. Later, Orwell found a similarly apt image for phrases that have a life of their own and that do your thinking for you. About one author, he said, "His words, like cavalry-horses answering the bugle, group themselves automatically into the familiar dreary pattern" (*Collected Essays*, 349).

Surely it is not merely amusing that this language is international. *Plastics Words* was written between 1985 and 1987, before the collapse of communism in Moscow and, incidentally, also before the linguistic contours of the European Union (EC) were clearly recognizable. In those years, when my daughter asked if I believed that the Berlin Wall would someday be torn down and the two Germanys would again be united, I responded that even if it did happen we would never experience it. Nearly everyone thought that then. But in Chapter 1 of this book I maintain that the fundamental concepts underlying the different political systems can be seen as similar if one exchanges some terms: that "sexuality" in West Germany and "development" in East Germany had nearly the same meaning, and that it was only a matter of time—in those days, one was somewhat prudish in Eastern Europe—until "sexuality," in its current meaning that says everything and means nothing, would arrive there. It is alarming how identical the fundamental concepts have turned out to be. Through the common basic building blocks of modular plastic words, the tracks were laid for the sudden and often shockingly brutal way that East Germany was remodeled and hitched like a caboose to our system. After "die Wende," which literally means "the turning" and is what most people call reunification, someone sent me a talk by the mayor of a small East German town, who until then had spoken perfect party-Chinese and now had excellently mastered the abracadabra of plastic words. In a weekly magazine, I even read the success story of a family in another small town, which had quickly risen to the heights of being the largest mail-order distributor of sex-paraphernalia in Eastern Europe.

Perhaps I should explain why many of the observations in this book are local in origin. While I was working on the analysis of plastic words, I sniffed the air around me for appropriate examples. It occurred to me then that in Freiburg—the city where I live—I could hear meetings breaking up all around me, and the same thirty people were emerging from their meetings spouting words that seemed to float in the atmosphere like a contagious cloud. "My" words clearly were the carriers in this outbreak. The minister-president of Baden-Württemberg, the state in which Freiburg lies, came and told us that our city was impoverished because it lay in the shadow of the French border and needed a "structure-leap" (*"Struktursprung"*); the senior mayor of Freiburg repeated after him that we needed a "structure-change" (*"Strukturwandel"*). A series of articles appeared in our local paper under the heading "Freiburg 2000."

These articles signaled, on the one hand, the road to hell of the "structure-crisis" ("*Strukturkrise*") and, on the other, the heaven of "structural-development" ("*strukturellen Entwicklung*"). I became acquainted with a development plan for Freiburg (I understand that these things are called "strategic plans" in the United States) that, despite its bureaucratic language, had an eccentric effect on me, and I wrote a satire with the title, "How Does One Turn a City into a Laboratory?" Therein I portrayed a word-generating machine that looked like the one that Jonathan Swift's eternal Gulliver saw in the Great Academy of Lagado on his travels. In my satire, the machine stood in the city hall. With a light turn of a knob, one could produce endless new development plans, using different combinations of thirty or forty words. Later, the satire appeared in my novel *Schauinsland*, whose protagonist is our city, where it found friends and foes.

Freiburg is a medium-sized city in southwestern Germany, two hours by automobile south of Heidelberg, where the hills of the Black Forest open themselves to the flats of the Rhine. In its center stands a stately Gothic cathedral with "the most beautiful tower in Christendom," as Jakob Burckhardt put it. Here, Berthold Schwarz invented gunpowder. Influential scholars have belonged to the University since the sixteenth century. Max Weber was here for a time, as were Husserl, the polymer-physicist Staudinger, and Heidegger. Freiburg is a city of musicians, only slightly industrialized by pharmaceutical, chemical, and other industries, which lies in a wine-making region and is bordered by vineyards. This is the city that gladly does itself in with its plans, goes deeply into debt, sees this as normal and obviously needed, and does all this in hopes of attracting new, rich settlers. With plastic words, the planners convince the citizens that they need the "improvements." The wealthy capital of the state, Stuttgart, insists on "development." In the city, there is resistance. The plans change. At one time, some wanted to conjure a new industrial center out of nothing, a total restructuring. Now some would like to make Freiburg a convention center. Freiburg sees itself as the "ecological capital of the country" and simultaneously works on plans for an autobahn, which would run right through the middle of the city and then through the Black Forest, which would lead to a bridge between Western and Eastern Europe.

Near Freiburg is the Kaiserstuhl, a volcanic range where wine grapes have been cultivated since Roman times. Here, not too long ago, in full sight of the entire German public, the Stuttgart government's plan to build an atomic power plant was successfully foiled by protests. At the same time, with barely any opposition, plans were carried out to turn the entire region into an artificially terraced landscape, which has resulted in overproduction and, hence, a soured landscape and sour wine. In brief: Today we find ourselves on an earth overrun with "programs" that have arisen everywhere out of a game plan, like the directions on a child's box of Lego blocks.

This book is neither pessimistic nor optimistic; it wants to say, "Stop transmitting messages of NOTHING." We settlers need time and place to criticize, debate, and decide. This is clear from any glance at the living chaos of our beautiful Freiburg. The best we can hope for is that this book will help to end the injustice.

I am delighted that this small essay is now appearing in the United States and that it is in the care of the Penn State Press. It has clearly proven to be a difficult book to translate into American English. If the translation works, naturally and without need of interpretation, it is thanks to Jutta Mason's and David Cayley's insightful and painstakingly thoughtful work. In those places where the German original was particularly difficult, or where minor changes were needed for English-speaking readers, the final form has benefited from the extraordinarily prudent and careful editorial work that was entrusted to Peter J. Potter, Andrew Lewis, and Judith Van Herik. I heartily thank all three. Further, I thank Professor Judith Van Herik for translating this preface and for suggesting the book's splendid new subtitle: "The Tyranny of a Modular Language."

# Preface

In this essay I try to describe how vernacular language has recently been transformed and what the consequences of this transformation are. I want to begin with a short explanation of some of the concepts that appear in the title and in the chapter headings.

The transformation of the vernacular can be grasped by studying a small group of words and a way of using them that is characteristic of the last few decades. I call them *plastic words*. Linguists might call them *connotative stereotypes*. Everyday use of the word "communication" is an example. These nouveau-riche nephews of science which can now be found in the vernacular total hardly more than three dozen.

When I say *the vernacular* I mean the commonly used everyday speech that forms most of public discussion as well as our private conversation. I distinguish this common language from the professional language of the sciences. "Ordinary language" or "colloquial language" or even "mother tongue" could also convey something of the broad sense I want to give to the term "vernacular."

In the final chapter I speak of the "mathematization" of the vernacular. By this I do *not* mean that the vernacular is actually moving closer to the language of mathematics, in the sense of becoming ever more precise and transparently logical. Rather, the expression points to a dangerous "mingling of spheres." I want to draw attention to the penetration of the everyday world and its language by elements and principles that have some mathematical properties. When I tried to describe the usage of plastic words in the vernacular, I noticed the curious fact that I could use some of the same terms to describe mathematical concepts. My surprising observation that plastic words have, in some ways, the abstract and ahistorical character of numbers is the basis for my concept "the mathematization of the vernacular." It describes a diminished version of the long-observed "scientization of the vernacular." That phrase is not usually meant to imply that the vernacular is becoming scientific, but rather that it has been colonized or invaded by science. Science is totally altered in a vernacular context. It becomes contradictory, doctrinaire, and imperialistic. Just as mathematics is deformed when it appears in the vernacular, so is vernacular language deformed when it is mathematized. This mingling and reciprocal deformation is my theme. Insofar as it affects language it affects worlds. I will not try to cover this reciprocal confusion of spheres by using the common prefix "pseudo" to speak of pseudoscience and pseudomathematics, although these terms would apply. Instead I will seek the still unrevealed connection between the dominant scientific ideas and their caricatures in everyday speech.

This book would not have been written without the inspiring and animated conversations I had with Ivan Illich, which began in 1981. The protestations of Freimut Duve were useful. The book also owes much to my conversations with Barbara Duden, Gustavo Esteva, Jean Robert, Wolfgang Sachs, José Maria Sbert, Ludolf Kuchenbuch, Peter Oberliessen, Jürgen Schiewe, and Gunhild Poerksen. I do not claim right of ownership over the messages in this book.

# Introduction

Plastic words are not new in how they look but in how they are used. They have been fashioned for the purpose of laying down the tracks and outlining the routes of a civilization that is covering the globe with gathering speed. Their origins can no longer be discerned. They resemble one another. It is as though there were a place somewhere in the world where these words were being released at intervals, as though at an unknown place there existed a factory releasing them complete from its assembly line, or as if they were coming into being simultaneously in many different places.

They may not be noticed, but they are present everywhere: in the speeches of politicians and on the drawing boards of city planners, at academic conferences, and in the ever more taken-for-granted in-between world of the media. They invade private conversation. When they first appear, they are fashionable and command attention; but then they merge with the everyday and soon seem commonsense.

In the spring of 1985 I attended a conference in the little Mexican mountain city of Tepotzlan, involving several notable industrialists, politicians, and academics of that country. The discussion was about how Mexico could take advantage of the most recent developments in high technology. Please note: not *whether* this ought to be done, but *how*. A professor and banker from the United States opened the meeting and explained his "philosophy." "Man" has two especially noticeable attributes, he said. He devises techniques and he organizes society. Between these two types of activities there is an "interplay" and they mutually uplift one another. . . . After "man" had been reduced in this way to these arid attributes, the conference was able to begin. The discussion was dominated by a number of words that floated through it like driftwood: "progreso," "proceso," "modernización," "necesidades," "comunicación," "información," "crisis," "desarrollo." The North American expressed himself a little differently. He replaced "desarrollo" with "development," and was ahead of the others in that he seemed to be already settled on the high plateau of "high tech," whereas they had to orient themselves toward the shining mountaintop of the future by using his position as a marker. Otherwise, there was unanimity. The words made this almost boundless consensus possible and dispensed with the question: "Why?"

I was only an accidental guest at this meeting; and, during a break, I remarked to a Mexican friend that the talk seemed to consist of no more than a hundred words. My friend shook his head and said quickly: "With a hundred words you could become president! Here there are barely fifty."

Or perhaps there were only fifteen. Fifteen intercontinental words: enough seemingly to make a beautiful and diverse land subservient to a new technology.

The domain of language is as rich and various as nature itself. Recognizing that language and dialect are barely distinguishable, Barbara Grimes has calculated that 5103 languages cover the earth. But when one asks what the distribution is, an astonishing pyramid becomes evident. (The numbers may be only approximate because of the difficulties of definition, but they are still interesting.)

Asia and Africa each speak 30 percent of the world's living languages, the Pacific region 20 percent, and the American continent 16 percent. Europe, by contrast, has barely one percent—sixty-seven languages. In the birthplace of the nation state, therefore, the number of minorities is the smallest; the number of languages is barely double the number of states. Can we conclude from this that the nation state extinguishes languages? In every case history shows us that the modern state pushes them to the margins.

A second pyramid, presented by Florian Coulmas, is no less significant. (Here also we must remember the reservations I mentioned above.) Five languages—Chinese, English, Spanish, Russian, and Hindi—are distributed among 45 percent of the world population. Twelve languages are distributed among 60 percent of the world's population. In addition to those mentioned above, these are German, Japanese, Arabic, Bengali, Portuguese, French, and Italian. A hundred languages are distributed among 95 percent of the world population. Five thousand languages are distributed among 5 percent of the world population.

In Africa, for example, the widely distributed European colonial languages overlie a colorful, tightly knotted, and asymmetrical carpet of local languages. In Nigeria over 400 languages have been counted. For India Coulmas has counted 1652, among which there are several languages spoken by barely a hundred people today. Yet in Nigeria and India English is now the predominant language of official discourse.

The same thing seems to be happening in the domain of language as in other natural domains: a reduction of diversity. Monocultures are gaining ascendancy and overwhelming the globe. Wherever we look, fewer kinds in ever fewer variants of corn and rice and wheat; Chinese, Russian, and English; and sheep, cattle, and pigs look back at us.

The nation state weeds out languages. It is the salesman of global unification. If we continue to try to characterize the world atlas of languages in numbers, then yet a third series is significant. In the year 1945 the Charter of the United Nations was signed by 52 states; now the U.N. has 160 member states. Two-thirds of today's nations are therefore barely one generation old and see it as their task, insofar as they follow the European example, to put in place a single idiom as a unifying national symbol, to alphabetize this one language, to standardize it, and to "develop" it so extensively that it can "transport" and "communicate" the content of world civilization. They have it as their task to introduce an imported and fundamentally absurd idea of language and to accomplish in a few decades a "modernization" that in Europe took centuries, the separation of the vernacular languages from the universal Latin written culture of the Middle Ages. How could this be done other than by a radical simplification of what is enclosed within their boundaries? And how could it fail to leave the young states and their suppressed languages completely defenseless?

Decolonization since 1945 has actually reduced linguistic diversity. The more decisively Third World countries take over the program of "national identity" from Europe, the faster their forest of languages is eroded. Colonialism, it has turned out, is not a peculiar tendency of the European nation state, but a property of the nation state as such.

Five languages cover almost half the earth, a hundred languages almost all of it. The universalist orientation of the nation state destroys the diversity of living languages. But even these triumphant languages are not the peak of the linguistic pyramid. The peak is comprised of that small and spreading international vocabulary of a hundred, or fifty, or fifteen words with which the conference in Tepotzlan was occupied. I have already used several of them: "identity," "development," "transportation," "modernization," "communication." The nation state is the undefended gate of entry of these universal signs. By admitting them it becomes, paradoxically, the engineer of its own disappearance. These signs are our theme. They appear unremarkable. They are not "catchphrases" or even "slogans," "empty phrases," or "buzz words," but words that have entered much more quietly and become common sense. They are the everyday prison of perception: "Energy." "Sexuality." "Information."

One could call these words the master key to the everyday. They are handy, and they open doors to enormous rooms. They infiltrate entire fields of reality, and they reorder that reality in their own image.

For the most part, these words already existed two or three hundred years ago; but they have changed their meaning. The change was imperceptible, for "energy" or "information" still sound the same; they have been altered only subtly. Since science abandoned Latin and began to use the folk languages, Spanish and Italian and French, Dutch, German, and Swedish—something which happened only a short time ago, after all—it has drawn into itself concepts from these languages, altered them, and then released them in their new form back into the common language, where they then had enormous effects. With the word "information" this happened only twenty years ago.

I have studied this phenomenon in various ways. I have traced the emigration of words into science and their return into the common language in the work of Goethe, Darwin, and Freud. The pattern we can see in the past is an excellent key to the study of the present, as long as a clear distinction is maintained between scientific language and the language of everyday life, as two separate spheres. Popular concepts from the vernacular are transmitted into science or some other higher sphere, where they pick up the semblance of generally applicable truths. Then they wander back, authorized and canonized, into the vernacular, where they become dominant myths and overshadow everyday life. This happened around 1800 to "health" and "development," and then around the middle of that century to "the struggle for existence" and "natural selection." The two domains are united by a seemingly common language. The original transmission leads to the reverse transmission. And the scientific teachings of Marx and Freud reappear in the everyday as doctrines and myths that disable the vernacular.

It was on this theme that I fell into conversation with Ivan Illich in 1981, during a year when we were both at the Berlin Wissenschaftskolleg. The topic has not left me since. He suggested, or rather he urged me, to carry my research right into the present and to zero in on how the ordinary language of today has been transformed by science.

I took on this project only reluctantly. Describing the disabling of the vernacular has something depressing about it. It has not always been possible to approach it without breaking into a sweat and feeling dizzy. The returning immigrants from science form the outlines of a worldview that dominates everyday life. Can it really be that this worldview is built up as primitively as it appears to be? When one is occupied with the key words of the everyday, one is easily attacked by a feeling of emptiness— and also of hubris.

It is a terrible theme, but it cannot be ignored. The scientific penetration of the everyday and its language has increased by leaps and bounds in the last few decades, and each small sentence made up of scientific-sounding words spreads out over the whole of the industrialized world.

Amorphous plastic words are the elemental building blocks of the industrial state. These ciphers clear the way for operations on a grand scale. Wherever they are in place, everything is made ready for a smooth and unhindered transition. Political systems are almost irrelevant in the face of these universal signs. They spread out on both sides of the Elbe, the river that marked the boundary between the east and the west, socialist and capitalist Germany. Both states had denounced the sentimental agrarian idyll of the Nazis, and this denunciation hid the extent to which they retained the Nazis' technical pathos. Without hesitation they incorporated the legacy of the Nazi autobahn-builders of the 1930s.

For the Roman historians Tacitus and Pliny, the Elbe was the "Albis," the white river. It could not be called that today. The Elbe, like the Rhine and the Danube, has become the sump of the German industrial state. It is no accident that this pollution coincides with the spread of plastic words.

Are the words at fault? If one looks only at the words, they sometimes appear to be a skeleton that displays the structure of the world more clearly than a full ideological presentation would. But the words do not just make this structure visible; they are also agents. Language is a partially autonomous power that shapes reality as well as reflecting it, and these two functions interact. Countless diffuse expressions are squeezed into one concept and fastened onto one name, and this name gains a certain independence. One forgets that it contains only a bounded view of things; one confuses the name with the thing itself. The name achieves the inertia of an established institution. Words are channels that run ahead of history, and history follows them. They should be questioned constantly; but out of complacency, fear, or plain stupidity we allow them, again and again, to lead us. We tailor the language and then we behave as though we are in uniform. Sometimes it seems to me that if anything is guilty, then by now it is only the words. But as I suggest in several ways below, their guilt depends on where and how they are used. When used in the way I describe in this book, the words are modules of a new reality—a reality that locks us in a conceptual prison. This prison must be held to account. We must analyze the language we use. All denunciation of individual words, without careful linguistic analysis, amounts to nothing more than theatrical posturing.

Language crystallizes consciousness and forms an intermediate world. The romantics enthusiastically regarded it as an expression of the *Volksgeist,* the spirit of the people. I see it as an intermediate world in the more sober sense that it institutionalizes and sanctions social and historical practice. This was driven home to me when, by chance, I happened on some regional newspapers from the 1930s. Reading them I felt as if a smog of unreal words was floating above my head. The newspapers reported the speeches of statesmen and politicians. Words such as "fatherland," "homeland," "loyalty," "honor," "the people" (*das Volk*), "freedom," and "heroism" then still sounded traditional and honorable. Now, because of their use by the Nazis, they sound as empty as "communication" or, worse, ridiculous, or even brutal. To my ear, plastic words heard now are in one respect comparable: they sound friendly, smooth, positive, and consensual, but, while not in themselves evil, they mask brutality. With a word such as "development," one can ruin an entire region.

Language is an ever more invasive phantom world, simplified and ordered by humans, which is jammed in between the human being and the world. By entrenching ideas and their influence, language makes it possible for us to continue to live in insane systems. The mere fact that language develops reality doesn't make reality unreal. Even phantoms can become real; but then, of course, reality becomes ghostlike.

Can these words be examined? Can their properties, their characteristics, be recognized and named? Can a general outline of meaning be discerned?

They are amoebas, these things that are spreading out into our ordinary language. In our discussions Ivan Illich called them *amoeba words*. We are familiar with these transparent and only vaguely contoured little animals, when we see them under the microscope, gliding, apparently enlarging and then changing; they are imperceptibly slow, almost a nothing. The variety that was discovered first (in the eighteenth century) was named "proteus," and these humble little creatures, which multiply by division, can also attribute their name "amoeba" to the constant alteration of form that happens when they move. They move forward in a flowing motion, as though they were running on changeable little feet or "pseudopods"—changeable little creatures on illusory feet. So calling plastic words amoebas has something fitting about it. It sticks. But friends who are biologists have asked me not to associate those innocent creatures from the earliest times with these recent verbal monstrosities.

I decided to follow the suggestion of Thomas Weck, my German editor, and use the term *plastic words* in this project. By adopting this term I hope to suggest a stereotype's endless ability to generate new forms. From time to time I use the expression *amorphous plastic words*. In *Mythologies* (first published in Paris in 1957) Roland Barthes described plastic as an alchemical substance and expressed amazement at its total malleability: "On the one hand there is the raw material, on the other the finished object. Between these two extremes there is nothing; nothing except a channel that can be traced back to an employee with a visored cap, half god and half robot, who is supervising the process. Plastic is less a substance than an idea of limitless transformation."

A name is not a definition but only a handle that memory can grasp; it can orient us only partially. Plastic words have weak contours. But novelties can sometimes be grasped with an image. These words resemble the floats that support a fishing net. They appear separate, but beneath the water they are joined by the cords that pass back and forth between knots and together comprise a net. This net ties together our perception of the world and holds it captive.

Since plastic words either begin in the sphere of science or pass through that sphere, they can also be described as a kind of *scientific rubble*. Or they can be imagined as *bridgeheads of science* in the vernacular. Looked at in this way the entry of scientific words into the vernacular appears to be a form of *colonization*.

Words have *auras*. In her work on connotation, Beatriz Garza Cuaron compares denotation, that is, the designation of a thing, with the first wave that forms when a stone falls in the water; connotation, or the feelings, associations, and valuations the thing evokes, she compares with all the following waves. Plastic words seem to be composed only of the ring-like connotations, which move outward from wave 2 to infinity. The stone and the first wave have disappeared.

These images still leave us without a definitive reference point by which we can identify these words. Eventually it became clear to me that this is their character. They cannot be more closely delineated. Their precise meanings cannot be discerned. This is not in itself unusual. Vernacular words also have vague, abstract, and far-reaching meanings. They can refer to various aspects of an object field and have shifting boundaries. *But* as soon as I use them in a particular connection, I circumscribe their reference, and their significance becomes precise, concrete, and exact. Plastic words, by contrast, are rarely used in a particular, precise, appropriate manner. They are used as interchangeable modules. Because of this, they lose any potential for precision, concreteness, or exactitude.

"Love" for example is a word that by 1850 had encompassed elements that are quite remote from one another. According to its context it can mean a number of things, and it lies within the power of the speaker to use the word at any given moment in quite different ways. The word can take on a whole spectrum of meanings; it is expandable. The boundaries are fluid and they extend from the affection within a family, to physical love, to pleasure at a piece of music, to the love of humanity. Grimm's dictionary of German lists eight distinct meanings for *Liebe* and for each of these meanings a series of nuances. So the word has a wide range and a richly endowed content, different in every instance.

This semantic pliancy of words makes them remarkably suitable for targeting the multiplicity of reality in its many half-steps. One is generally not in a position to define a concept offhand. Yet in practical usage, words take on special meanings according to the place assigned to them by the speaker, and so are nuanced according to his wishes.

This usual course of events, recognized in linguistics, does not seem to apply when it comes to using words like "communication" or "sexuality." In these cases, the speaker has apparently lost the power of definition. And this is in fact the first criterion by which we can recognize plastic words.

In Chapter 1 of this book I elaborate further criteria that permit us to mark out a class and to identify the words belonging to it. There I sketch a composite image of these words, in a rigorously formal way, and identify three dozen of them. In Chapter 2, I ask whether the way these words appear to me to be used is really historically new, and where the newness lies. I examine the history of the word "information" and distinguish it from other classes of words: technical terms and abstractions, buzzwords and empty phrases, catchwords and slogans. In Chapter 3, I give a series of examples that demonstrate how these words are inserted as extremely plastic building blocks (or modules) in models of reality that then become available to planners. In Chapter 4, I discuss how the words, when they appear in the jargon of experts, conquer and destroy ordinary language. In Chapter 5, I introduce the bold expression "mathematization of ordinary language," which allows me to summarize my theory.

The rigorous analytical form is not only meant as parody; at this moment I see no other way of approaching this phenomenon.

# 1

## Plastic Words on Both Sides of the Elbe

## The Composite Image

I begin with examples drawn from the very different societies of East and West Germany before the collapse of the wall, examining the use of the word "sexuality" west of the Elbe, and of "development" to the east. The two examples show surprising similarities when one tries to grasp them in a more abstract context.

## Sexuality

This word arrived in the vernacular within the last three decades.

But it has not yet settled so firmly that a speaker who tries to use it can nuance it in his own way. For example, if he says that he "can't deal with his sexuality," then he obviously *has* something that is making a problem for him; it is a foreign object for him, something for which others have a responsibility. It remains independent of the context of usage: autonomous.

A speaker deals with this word as a finished building block, in the same way that a scientist deals with a term, but the arena is no longer science but vernacular speech.

In the late 1980s an autobiographical novel appeared in Germany that was much discussed at the time. In this book a woman confessed that the "Fairy Prince" she had married had since become a severe disappointment to her. She pronounced that in earlier times she could never have admitted her own sexuality but had now become strong enough to live out her sexual needs even in casual relationships, since she had learned to "deal with her sexuality." Already, "sexuality" had made its appearance as a fixed element, which cannot really be comprehended by the reader.

Such a word not only *appears* to be independent of its context, it actually *is*. It has been penetrated by the science of psychoanalysis and molded by it. The adjective "sexual" can be found in nineteenth-century dictionaries, in Joachim Heinrich Campe's dictionary of foreign words (1813), in the very rich and almost unknown *Greater Dictionary of the German Language* by Jacob Heinrich Kaltschmidt (1851), and also in Daniel Sanders's dictionary (1865). "Sexuality" is a word borrowed from science in order to enhance the prestige of the speaker. It is a word that has rolled down from a higher plateau.

How plastic words are used bears some resemblance to metaphor. In a metaphor a concept is carried from one quite distinct domain to another. In "the harbor of marriage," for example, the harbor is carried into the marriage. Metaphors are abbreviated comparisons that subsume separate domains.

To speak about "my sexuality" in the vernacular also implies a brief joining of two spheres that are separated by a chasm. The scientific expression belongs in the circle of the impersonal, the objective, and the general; the scientist tries to move in the direction of the generalizable. His language is not made for private, intimate speech. Wherever something is experienced as very personal and often also as unique, the flexible and nuanced vernacular is used. In the private sphere, consequently, there is a rightful aversion to technical terms. Their use always, or at least for a long time, retains something metaphorical. But this metaphorical character tends to be obscured by the fact that these words that are carried over have nothing picturesque about them. By nature, however, metaphors and comparisons are picturesque—even pictorial. The image of prowling that lies behind "cat burglar" is an obvious example. Plastic words show no trace of their origins.

In Freud's concept of "sexuality" there was originally a visual content. He took the psyche as an apparatus inside which measurable or at least estimatable quantities of energy circulate. This energy can be dispersed, repressed, displaced, heightened, or lessened. The idea of the psyche as an apparatus for energy distribution derived from physics, and at first constituted a revealing perspective. Freud favored the imagery of natural science in his psychology because he thought it came closest to representing the way things actually happen. He sketched the psyche from the point of view of natural science. Words derived from his work, and dependent on nineteenth-century concepts of physical energy, have long since been incorporated into ordinary language. Tension is "stored" and "released." These images have become so pervasive that their pictorial content has faded.

Their effect is all the stronger for that. "Sexuality" is a metaphor in a double sense, as a concept derived from science and by reason of the language of images from physics. The fact that it is a metaphor is hardly noticed any more.

The concept of sexuality has a wide range of meanings within psychoanalysis. After Freud's enlargement, it included virtually all aspects of desire or love that can occur in life, starting from the thumb-sucking of the infant. Freud himself names a whole spectrum of possible contents: "sexual love," "self-love," "love of parents," "love of the child," "friendship," "general love of humanity," and "giving oneself to concrete objects and abstract ideas," and then he summarizes: "Psychoanalysis . . . gives these love instincts [drives] the name of sexual instincts [drives], *a potiori* and by reason of their origin."

We do not need to deal here with the difficulties that came into the psychoanalytic school especially through the little phrase "*a potiori* and by reason of their origin." It will be enough to take a look at the universality of the concept.

Of course Freud was not able to introduce this enlarged concept of "sexuality" into the vernacular. But he still equipped it with far-ranging effects. The word seems to hover in a mist. On the one hand it has retained its older narrower meaning—one uses it to refer to the area of the genitals; on the other hand it has become a universal explanation. Now, even outside the psychoanalytic school, all varieties of love count as sexual because they all derive from basic submerged drives: as for example in the following title of a TV series: *Sexuality as a Part of Human Communication.*

At the same time the concept tends to flatten out the many levels of possible vernacular meaning documented in Grimm—connection, attraction, friendship, tenderness, love, yielding, passion. It crowds out this treasury of synonyms with all their nuances.

Meanings previously considered autonomous—"friendship," "brotherly love," "love of humanity"—lose their ability to stand on their own. The *verbum proprium*, the word which alone fits in a particular connection, loses its particular aptness and is at least weakened, if not replaced.

Where in earlier times there would have been silence, or an indirect form of address would have been chosen, it is now possible to have a sentence like this one: "Today I would be strong enough to realize my sexual needs even in a 'casual' relationship and to avoid any strictures." Such a sentence would have been unthinkable a generation ago. It illustrates an alteration of ordinary speech.

The word "sexuality" is cut off from the rich store of gesture, expression, and pantomime available to ordinary language; it is toneless.

The concept brings a gigantic area under one appellation. All trouble and woe can be treated at one point, to cite Goethe's Mephisto; all private and public well-being depends on genital satisfaction. Thus it sets up a universal form of expression that is diffuse and impoverished.

The actual object to which the word "sexuality" refers is already hardly graspable in Freud. "We do not possess a generally recognized indicator of the sexual nature of a process . . . ," he writes, and explains that such a criterion must be discovered in the area of biochemistry. In

*The Language of Psycho-Analysis* it says that "all that one can postulate in psychoanalysis is the existence of a sexual energy or libido, which can not be defined in the clinic, but which is revealed through its development and transformation." This postulate from natural science has arrived in the vernacular as an empty space, so to speak.

"Sexuality" is an energy that can only be postulated as a useful assumption from which one can work. In this it resembles the concepts of more recent physics. Might we not then give it the name of the Greek letter *epsilon*, in place of the grandfatherly and suggestive word "libido"?

The concept "sexuality" has no historical dimension; nothing in it points to a local and social context. It translates life stories into the terms of natural science and says that everything is basically the same.

Because of its encompassing generality the concept gives the impression that it fills a gap. It awakens and satisfies a need that did not formerly exist. A person in the Middle Ages had no "sexuality," neither the word nor the thing. This is a relatively recent construction.

The concept makes older words of relationship—attraction, friendship, love, passion—appear old and out of date, since it explains all of them as derived from *one* energy. In the light of the plastic words, the vernacular appears to be a rubble of outmoded ideas.

"Sexuality" is strong in connotation; instead of power of designation the word has an impressive aura. It suggests something positive, something that is both a property and a basic need. Whoever says "in earlier times I could never have brought my sexuality into such 'casual' relationships," appears to possess both a property and a particular insight into this property.

The word signals science. It silences. It makes known: *Roma locuta est.* An "I" who gives up the private quality of a relationship, its yielding, its indirection, or its immediacy, in order to speak in a possessive way of "my relationships," "my sexuality," "my overreaction," subsumes her experience under very general categories and connects herself to a common structure of thinking. She transforms the private sphere with an objective language that was intended for a completely different function and assumes a viewpoint distant from her own. She turns herself into an object of science. She becomes a case. In doing this, she at first gets relief, she is able to gain distance—but she also delivers herself up to science.

Such an "I" takes an assigned place. "Did you have your orgasm?" asks the mother in a novel by John Updike. The use of the possessive pronoun is significant. It shows paradoxically that one is under the control of others. Such a word—in fact the whole vocabulary being discussed here—is used less to impart something in ordinary language than to serve a function: in evenly subsuming experience into "science," it strengthens social hierarchy.

It builds a bridge to the world of the experts. The content of the word "sexuality" is only a nebulous white spot for us, but it hints at another world in which others know more about it. Knowledgeable persons exist who can teach us how to cope with this foreign body, which they administer. Such a word increases the need for expert help.

The concept is comparatively new; only in the last fifteen or twenty years has it entered into common (public or private) usage. In other countries, Mexico for example, it has hardly entered even yet. The source material for a new, comprehensive Mexican dictionary now in preparation contains only six instances of the word "sexuality." But the collection of this material was already completed in 1973, and today, so the director of the dictionary project, Fernando Lara, assures us, the number of lines in the entry would be very much higher.

In the area of word formation, the concept is very productive, similar to "health" around 1800: Sexual research, life, freedom, knowledge, education, atlas, enlightenment.

And finally, after some delay, we meet the word internationally.

The word "sexuality" therefore has a series of singular features. Is it actually possible that the usage of the word "development" on the other side of the Elbe before 1989 resembled the usage of "sexuality" on this side?

# Development

"The successful accomplishments of the writers of our region are in many ways bound up with the development of our workers' and farmers' state," it says in the introduction to a booklet called "The Writers of the Region of Erfurt" (1979).

"Development" is used here in a very general way. In the GDR it more or less took the place of "history." Perhaps one could say that it spoke of the history of the GDR in a particular way: it organized millions of possible pieces of historical data along one line: "development."

The word was endowed with a plus sign. The "development of our workers' and farmers' state" ran parallel to and corresponded with the "successful literary accomplishments of the writers in our district." It was a positive word.

It presented itself quite unobtrusively. The writer of the introduction could have left it out; it said very little and could almost be taken for granted. Development was an automatic attribute of all events, emphasized for its own sake. "As a result of historical developments the majority of the old, tradition-rich and productive German universities and high schools were located in the territory of West Germany," it said in a book about Jena (1980).

The vernacular word "development" had more explicit scientific overtones in the east than in West Germany. It was taken into science from the vernacular in the eighteenth century and rapidly spread to all disciplines there—a point we will return to. Then it emigrated back into the common language, carrying a new scientific definition. It enjoyed the authority of the reigning scientific teachings. Anyone who visited the Museum for Pre- and Early History in Weimar and read the explanations that accompanied the exhibits, or the brochure, would be impressed by their scientific tone:

> In this overview of the geological formation of the earth and the developmental history of plants and animals, the visitor will see a demonstration of the natural requirements for the development of the human and the formation of society. Valuable archaeological discoveries as well as dioramas, models, pictures, and graphic representations arranged in an easy-to-understand way will give an overview of prehistoric society and its development up until the formation of class society and the German feudal state.

The representation of these historical processes should also help students, young people, and adults to acquire and use basic philosophical ideas and lead them beyond the emotional experience into a better understanding of the connection between the development of nature and society.

In three sentences "development" is used four times, "formation" twice, and "process" once. Science pervaded the culture of East Germany even more thoroughly than it did the culture of the west. Scientific socialism was known to be the foundation of the state, and "development" one of its universal keys. Experts molded public discussion; and, wherever people talked, the words they used aligned them with a higher realm where these words were actually understood. The result was doublespeak, a two-tongued language in which ordinary talk and expert talk were intertwined. The language of the elite colonized and ruled the vernacular.

The word "development" was so common in the former GDR that it seemed to take on a life of its own. It did not acquire nuance depending on its context, but rather marched through the language as an independent authority. A speaker could not make it subservient to the structure of his sentence, but rather used it as a stereotype, a finished block, an object, which then seemed to lead the sentence in its own direction. One thinks in words. Syntax escaped control, and the process could be observed wherever the media let the echo of isolated words swell into a roar. Try to read the following sentence from a book about the Alma Mater Jenensis (1983): "As a result of the scientifically based orientation of the Tenth Party Day of the SED [Socialist Unity Party] on the tasks of the sciences as well as the role of the universities of the GDR, a secure prospect has opened up for the Friedrich-Schiller University involving a comprehensive development and more demanding challenges relating to the requirements and expectations of the socialist society."

In such statements one can hear no voice.

The *Dictionary of Marxist-Leninist Philosophy* (1986), for example, gives the official definition of "development": "Movement in an ascending line, the unification of quantitative and qualitative change by dialectical negation. . . . The material world forms a system of qualitatively distinct developmental stages, which are connected to one another in the history of development." The phrase "Movement in an ascending line" describes almost nothing, but it brings a whole world into *one* descriptive word. A diffuse image without nuance replaces precise description. Language grows thin and watery.

The ascending line aims at infinity. The word "development" names the property of forward movement. Everything that does not explicitly or actually correspond to it is hopelessly backward and out of date.

The original bold image in this word—in the seventeenth century in Germany one still "developed" scrolls, and from this image the concept of development was originally nourished—has long paled. The word no longer transmits an image; on the contrary, it eliminates images from the historical world. It is an apparently neutral abstraction that describes a process. This meaning is not yet very old. The original active and transitive verb does not become an intransitive verb until about 1800. "A develops B" now becomes "something develops." The concept "self-development" came into use at the same time. I can only hint at the history of the concept, which is fruitfully described by Wolfgang Wieland in the handbook *Basic Historical Concepts* (*Geschichtliche Grundbegriffe*). Around 1800 "development" changed from an action into an *alteration*, whose cause and object remain unnamed. These alterations—such is the nature of intransitive verbs—then acquire attributes of natural law and an inherent goal. Development appears to be a natural process.

Marx prepared the ground thoroughly for the usage of "development" in the GDR. It is true that he emphasizes (and Engels does this perhaps even more) that the developmental stages, from inorganic nature through organic nature all the way to humanity, are divided by sharp breaks and cannot simply be seen as equivalent to one another and cumulative in effect. But basically the concept of a developmental law was applied to all levels of nature. "All matter and all developmental stages are determined by definite general laws, which are recognized and reflected by Marxist dialectic," reads the relevant citation in the *Dictionary of Marxist-Leninist Philosophy*. The idea of history as a natural process is no longer merely vague and picturesque. It is now analyzed as a process that operates according to iron. laws. History has been systematically transformed into nature.

This idea has had an enormous influence on public discussion. And this is true in Germany in the west, as much as in the east. After it became known at the beginning of December 1986 that the chemical company Hoechst had poured chlorbenzol into the Rhine, a government legal expert spoke about a "rather dramatically worsening development." The criminal act of an industrial corporation was painted over as a natural event.

The positive and apparently harmless word "development" rewrites history, paints over controversy, and locates unpleasant happenings in nature. It propagates a single version of reality. "Development," in its many-faceted generality, creates consensus.

After the end of the war there were serious struggles at the University of Jena until the new powers took command, antagonistic conceptions were "eliminated," and the orientation of the teachers and students was "adapted" to the thought of Marxism-Leninism. One might have thought that the author of the book about Jena I am referring to was speaking about actions, but he continued: "Only the ongoing developments of the fifties gave the teachings of Marx and Engels the place due to them at the University . . . this process completed itself on the soil of the German Democratic Republic."

The process completed itself as automatically as a wheel turns: what happened appears as a natural process. The ongoing development itself emerges as a person, an active subject, which makes room for itself.

This unobtrusive grammatical alteration has great consequences. The relatively young intransitive, in this case history presented as development, tends to become a new transitive, to become an active subject. Development develops. A historical process, with any information about who is doing what to whom omitted, almost imperceptibly becomes an original cause, whose actions affect objects. So history is transplanted into nature a second time, and historical actions cease to be subject to ethical questions.

That can happen so gradually that one is unsure whether the word refers to an event or an action. When the mayor of Weimar comments that a large number of companies share in "the development of our city," is "development" still used intransitively here with the city as the developing subject, or is "development" used transitively with the city as the developed object? Is "the development of our city," expressed grammatically, a *genitivus subjectivus* or a *genitivus objectivus?*

There are unambiguous examples in which the transition to the *genitivus objectivus* is perfect, for example when the *Thueringer Tageblatt* on 11 December 1986 cites the "Blueprint of the Decision for the Further Development of Land Management, Forestry, and Agriculture," or writes: "In the time-frame of the new five-year plan, the region of Erfuhrt will develop into a center for high technology." Development is a natural process that can be forecast and engineered. Protean and oscillating in its applications, the word has become an amoeba as big as a jellyfish, bobbing along in the wake of science.

The word was a tool in the hands of the experts who obeyed the party leaders on the other side of the Elbe and who now obey the leaders of industry on this side. Bunches of bureaucrats hang on the word. It is an almost limitless resource. A concept such as "developed socialist society," which designated the supposedly temporary (but still enduring) precursor of communist society, was constantly reinterpreted in ever new releases by the party leaders, and constantly transformed into new five-year plans by the caste of experts.

Intensive international use of the word dates from after the Second World War when Harry S. Truman made the word "underdevelopment" a keystone in his foreign policy. Until 1940 the German dictionaries are astonishingly devoid of references to this key word; not until the six-volume Duden in 1977 and the Brockhaus-Wahrig of 1981 was it evident, in 58 and 66 lines respectively, how pervasive the word had become. The article by Wieland mentioned earlier also underscores how recently this word embarked on its dazzling public career.

The semantic outline of "development" and that of "sexuality" show many similarities. I could further strengthen my argument that we are dealing here with a new class of words by examining the words "energy," "progress," "communication," and "information." But first I want to pull together the characteristics I have worked out so far. These characteristics allow us to discern the general class to which these words belong, and to sketch a composite image of a plastic word. By now we have a catalog of thirty criteria (a more complete summary is printed in the back of the book):

A. 1. The speaker lacks the power to define the word.
2. The word is superficially related to scientific terms. It is a stereotype.
3. It has its origin in science.
4. It is carried over from one sphere into another, and is in that sense a metaphor.
5. It forms an unnoticed link between science and the everyday.

B. 6. It has a very broad application (lit. domain of use).
7. It displaces synonyms.
8. It replaces the conventional, precise word.
9. It replaces an indirect way of speaking or a silence.

C. 10. It condenses a huge field of experience in one expression.
11. It is impoverished in content.
12. Its imagery is vapid and diffuse.

D. 13. It is historically disembedded.
14. It transforms history into a laboratory.
15. It dispenses with the question of value.

E. 16. The "aura" and associations of the word dominate.
17. It names a property and contains the appearance of an insight.
18. It has more of a function than a content.

F. 19. As a scientific "idealization" of something limitless it uncovers and awakens needs.
20. Its "naturalness" strengthens this pull.
21. The resonance of the word is imperative.
22. It has multiple uses.

G. 23. Its use increases prestige.
24. It leads to silence.
25. It anchors the need for expert help in the vernacular and serves as a resource.
26. It forms new words and is a flexible instrument in the hands of experts.

H. 27. It makes previous words look out-of-date.

28. In this sense it is new.
29. It is an element of an international code.

I. 30. It lacks an intonation and cannot be replaced by pantomime or gesture.

This anatomical atlas can be simplified by arranging these attributes in blocks. A plastic word would then have the following essential characteristics:

A. It originates from science and resembles a building block. It is a stereotype.
B. It has an inclusive function and is a "key for everything."
C. It is a reductive concept, impoverished in content.
D. It grasps history as nature.
E. Connotation and function predominate.
F. It generates needs and uniformity.
G. It renders speech hierarchical and colonizes it, establishing an elite of experts and serving as their "resource."
H. It belongs to a still very recent international code.
I. It limits speech to words, shutting out expressive gesture.

So, do "development" and "sexuality" mean the same thing? It seems to me that they signify different things, but what they signify is less important than what they mean. And the meaning is the same. These are close relatives of the myths of everyday life described by Roland Barthes. They are idols, magical and empty.

These words are at the same time very handy; they are the building blocks of reality blueprints. As quick as a wink, and after only the simplest introduction, they can be used to erect a new "model" of the world with Lego blocks.

Perhaps we should see what would happen if our two words were switched?

With "sexuality" on the other side of the Elbe, we get into difficulties. Officially the GDR hardly had sexuality for a long time. The border kept it out. Sexuality is no key word in the dictionary of Marxist-Leninist philosophy and therefore is not yet capable of hatching an expert bureaucracy in a state founded on that philosophy.

And yet the distance between the two states was not as great as might first appear. "Sexuality" began to find its way across the Elbe a long time ago, and by the time the wall fell, it was already as subversive as jeans and well on its way to becoming a recognized resource. In Christoph Heins's story "Dragon Blood" the plastic word multiplies the gray splotches of his dreary realism. The state had already discovered the sex drive as an instrument of control of the "population"; the youth lexicon had recognized this key word since 1986, and there were offices for sexual counseling. The Soviet Union had also become a frontier of sexual development. Under the headline "More and More Marriages Break Up. USSR: Hardly Any Information on Sexual Matters," the Associated Press reported in July 1987 that a leading sociologist had pronounced on the "sexual illiteracy" of his countrymen in the magazine *Ogonyok*. This strongly suggests that by the year 2000, the citizens of the Soviet Union would have had to be refashioned into a people possessing and standing in need of sexuality.

With "development" on this side of the Elbe, we have no problems. The Federal Republic has been overwhelmed throughout its history by waves of developmental mania and the urge to demolish whatever stands in the way.

In Freiburg we have been experiencing a milder form of this mania since about 1985. The minister-president of Baden-Württemberg and the mayor of the city agreed at that time that because the city lay in the shadow of the French border, it needed a "structural leap" if it didn't want to miss the gravy train. Since then there has been no end to the propaganda in favor of turning Freiburg into a launching pad. I am not concerned with speaking for or against this propaganda, but with the language in which the "development" of the city, as *genitivus objectivus*, is planned and a historical place transformed into a laboratory. But before we can undertake a description of this language, it is necessary to expand the list of the plastic words and to flesh out my portrait of this word type more clearly.

I have already sketched a composite image and set up a catalog of criteria. Do these criteria allow us to identify a plastic word?

"Love" clearly does not qualify. The word hardly fits the criteria in a single place.

"Relationship" is quite a different case. In the last fifteen years, this word has had a career that makes me suspect that it may belong in our class of words. And in fact it fulfills almost all the criteria: it is scientific, universal in application, and has the reduced content, the ahistorical naturalness, and the positive resonance and authority frozen in a function. The word awakens needs and constructs a bridge to the world of human relations experts. It appears to fit into the new, freely usable, international code.

The word "classification" fulfills one criterion but not the others. It is a scientific term but the speaker doesn't necessarily lack the power of definition. The scope with which it can be used is limited. It does not replace its synonyms. Nor does it reduce its field of influence to a single word. Its "object" is clearly graspable. . . .

The word "classification" can be left off our list.

"Communication," on the other hand, once again fulfills all the criteria. One can check its current usage against the list of characteristics.

It is therefore possible to identify words that belong to this new classification. "Sexuality," "development," "relationship," and "communication" belong there; "love" doesn't fit; and "classification" is disqualified in important respects. The composite image of a plastic word serves as a tool that allows us to recognize other such words. The whole list comprises perhaps not more than thirty or forty words. Provisionally, the following belong there, or at least look highly suspicious:

| | | |
|---|---|---|
| basic need | identity | resource |
| care | information | role |
| center | living standard | service |
| communication | management | sexuality |
| consumption | model | solution |

| contact | modernization | strategy |
|---|---|---|
| decision | partner | structure |
| development | planning | substance |
| education | problem | system |
| energy | process | trend |
| exchange | production | value |
| factor | progress | welfare |
| function | project | work |
| future | raw material | |
| growth | relationship | |

This list is neither fixed nor stable. Several words are questionable. Do the words "decision," "exchange," "role," and "value" belong here? We must not "take language too literally," according to Werner Heisenberg. There are degrees of belonging; almost all the words listed here belong to the hard core. Others are perhaps only on their way to this status. The invisible factory of the plastic words draws on suitable verbal raw material, refashions it, breathes new life into it, and periodically releases new batches of words. The effects radiate into the language.

When I say that the words listed above almost all belong to the hard core, I want to confront another likely misunderstanding. Words are not supposed to be stigmatized here; I want to expose a way of using particular words that is typical of our time. The words "development," "sexuality," and "communication" are in certain places as correct, clear, and precisely usable as other abstractions. Here we are discussing a particular usage of these words. If we imagine them as dice, we are looking at one side of them. We live within power relations that tend to foster this aspect of their meaning. Following our image, these power relations ensure that the dice almost always come to rest with the side we are concerned with face up. The meaning of words varies with their context, as was explained in the introduction; the present historical context evidently favors the connotative stereotypes I have described as amorphous, amoeba, or plastic.

Dictatorships have the capacity to order and enforce certain usages; there is no more effective instrument of propaganda. They proceed as one must proceed if one wants to gain the use of a word: they determine the connection in which it must be used. They order the context of the verbal sign, they determine whether it ought to be outfitted with a permanent plus or a minus. They have a tendency to dualism and usually a large fund of plus and minus signs. We know of a memo from the Nazi time, from the Ministry of Propaganda to the Newspaper Administration, ordering them to use the word "fanatic" only in a positive sense.

The structure of our society prevents the determination of language usage by government edict. But there are various ways in which the typical usage "comes about." For the variant that we are talking about, an overpowering context is set up in many different places simultaneously. And this context is precisely what leads to the domination of this variant. So everyday language alters itself imperceptibly, since the vocabulary generally exists already and the words keep the same sound. Language criticism must address language as it is *used*, as a fleshed-out social norm — not just as a "system." It follows that whoever wants to control language use had best intervene in the existing usage and try to make the desired context normative.

But what I have been calling "usage" is not easily distinguished from "system." The distinction is artificial. The usage of a word is its meaning; variants in use are variants in meaning. It ought to be possible and will probably be necessary, when words that float between science and everyday speech are entered into dictionaries in the future, to note the nature of their usage in ordinary language in italics. Dictionaries note in italics that such and such a meaning is *archaic*, another is *slang*, another applies only in some specialized field. Perhaps they should also note *plastic* as a category of usage and name some of the characteristics I have described, for example, that the speaker cannot give them a determinate or specific meaning, that their range of use enlarges to universality, that they squeeze out other possibilities, and that their power of denotation is diminished in favor of a strong connotation or "aura."

So we are calling a limited number of words "plastic words." They are distinguished by a typical variance in meaning that is effective within our society at this moment. When one looks at them in this way it becomes evident that the *words* should not be branded. I am not proposing to imitate the purist who hunts down foreign words. Even when one speaks of "slogans," "clichés," and "buzzwords," what is meant is a typical way of using words that could in a different context be used quite differently.

Our theme then is the most recent phase in the standardization of ordinary language. After the nation state has unified and standardized the languages of its own territory, they are then reduced to a small code capable of international circulation. I suspect that in Germany the class of words with which we are dealing here transformed our German vernacular more thoroughly than did our numerous borrowings from the Americans. I also suspect that they require quite a different kind of attention than our often useful foreign borrowings, whether old or new.

But do plastic words produce their effects as words or actually more as concepts? Do we not have to decide whether we're talking about words that spread out into the vernacular, or the concepts that form their background; about a vocabulary that provides a *leitmotif*, or about the modern concepts that lie behind this vocabulary? In the language of linguistics, are we beginning with the word (semantically) or with the object (onomastically)?

We must keep both in view. Plastic words are first and foremost *concepts*. "Development" is primarily a concept, or, to use an expression from the natural science of Goethe and Lichtenberg, a "way of imagining." In the nineteenth century, "development" was a particular way of representing the common element in a variety of natural and historical phenomena. They could also be represented differently, for example by the word "evolution." The primary meaning of the concept of development since the 1950s can also be expressed with other words. It is anchored in a hundred different ways in our ordinary language by supporting expressions. Such nouns as "improvement," "growth," "modernization," and "innovation"; such verbs as "improving" and "modernizing"; the adjectives "overhauled," "progressive," "modern," and "out-of-date" all support "development."

On the other hand we must pay just as much attention to the *word*. "Development" also works through its name; the sound itself conjures many meanings and associations. It is a pointed abbreviation, a sorter of diffuse data with a particular ring to it. Such an isolated name can, as I mentioned at the beginning, take on a life of its own.

The final objection can now be dealt with. "One could argue that two and a half dozen of such 'amoeba words' could be found in every epoch," the *Frankfurter Allgemeine* wrote on 20 May 1986 in response to a presentation of this thesis. Could one really argue this? The question deserves clarification. Is this type of word new? Does it illuminate a new phase in history? What distinguishes this type from basic concepts of other epochs, from scientific terms or mere abstractions, and finally from the darlings of the language critics: "buzzwords," "empty formulas," "clichés," and "slogans"? In order to get at these questions we will look at the history of the word "information."

# 2

## Are Plastic Words a New Class of Words?

The following diagnosis of a crisis in the language dates from the year 1875:

Everywhere *language* has fallen ill, and the oppression of this dreadful sickness weighs on all of human development. Language has continually had to climb up to the highest level it could reach, in order to grasp the domain of thought, and has therefore had to move as far as can be from its profound impulse simply to correspond with things as they are. Thus, in the short space of contemporary civilization its strength has been exhausted by this excessive effort. It can no longer accomplish precisely that purpose for which alone it exists: to enable suffering people to understand one another's most basic troubles. Man is no longer recognizable in language. He can no longer give a true representation of himself. In this dimly intuited condition, language has everywhere become a power unto itself, which now grabs people with ghostly arms and forces them into places where they don't even want to go. As soon as they try to understand one another, and to come to an agreement about some work, they are seized by the madness of general concepts. The very sounds of the words enchant them. As a consequence of this inability to make themselves known, whatever people create together carries the sign of their lack of mutual understanding. It corresponds only to the hollowness of these tyrannical words and concepts and not to man's actual troubles. So to all its other sufferings humanity must add this new suffering: that words lead to actions which no longer correspond with feelings.

This diagnosis by Nietzsche in his *Untimely Meditations* (IV, 5) seems to me to go further than the famous Chandos letter of Hofmannsthal, which is often cited as the beginning of "modernity." The author of this passage thinks historically and reveals a historical moment. Nietzsche holds a centuries-old overextension of language responsible for the independence language has assumed. "It spins the individual into the net of 'obvious concepts'," he goes on to say, "and teaches him to think correctly."

In hindsight this passage sounds prophetic. But if one tries to understand in what way it is prophetic, then it "only" seems to show a clear and sober outline of our contemporary circumstances. The chasm of which Nietzsche speaks has grown deeper, the distance between the individual and the ghostly arms of the universal concepts has grown greater. If one wanted to tell the story of how language climbed to "the highest level it could reach," one would have to begin with the last years of the Middle Ages. One would have to sketch in the end of the Latin written culture of the Middle Ages, which endured for almost a thousand years, and describe how the European vernaculars became the vehicle for scientific, philosophical, religious, and political ideas. This involved a gradual elaboration, violently contested, of a common language of education. In the new nation states of the Third World, vernacular languages are now undergoing this transformation in the space of decades. Regrettably, how the sciences moved from Latin into the vernacular is a subject that has hardly been addressed until now. In Germany the change took three hundred years, beginning in the fifteenth century and ending in the eighteenth century. And, as it happened, a new complaint appeared, for example in the *Patriotic Fantasies* of Justus Möser. A scholastic "refinement of concepts" Möser said, had overburdened the vernacular with impenetrable scientific coinage, so that one person could no longer understand the next.

By the middle of the eighteenth century, around the time German gained universal acceptance as the common language of education in the region, and perhaps because of it, political and social concepts also began to change. In the five-volume series *Basic Historical Concepts (Geschichtliche Grundbegriffe)* this transition is called a "Sattelzeit," literally a saddle-period. This work records the history of the social and political concepts to which the modern world owes its existence. Among them are ideas like "citizen," "democracy," "reform," and "state," but also "need," "development," and "progress." Reinhardt Koselleck notes in the introduction that there has been a "fundamental alteration in meaning of classical topoi" in the direction of "increasing degrees of abstraction." Nietzsche's diagnosis appears to have been confirmed. Koselleck's work concentrates on the historical epoch in which those changes occurred.

In the Jacobin phase of the French revolution, we can recognize a precursor of the global uniformity that is one of the themes of this work. Brigitte Schlieben-Lange draws it to our attention: "uniformité" becomes at this time the key category of French language politics. Making the language uniform was part of an attempt to achieve uniformity in general. She analyzes it, with the help of Rudolf zur Lippe's clarifying concept of "geometrization." Uniformity as a constructive principle of state policy is an ancestor of the connotative stereotypes still expanding around us, if any more expansion is possible. Schlieben-Lange notes the origin of "uniformité" in the domain of natural science and mathematics. In his *On Truth and Lie in the Extra-Moral Sense*, Nietzsche says: "Whereas metaphor avoids typification by the individuality and uniqueness of its pictorial content, concepts show the impressive structure and stiff regularity of a Roman column, and their logic suggests the kind of strength and coolness proper to mathematics."

Could it be that another root of the transformation is democracy?
Jürgen Schiewe cites Tocqueville. In chapter 16 of his work on American democracy ("How American Democracy Has Altered the English
Language," 1835) Tocqueville ascribes to the new form of government
an increase in three features of the English language: abstraction, personification, and blurring:

> Democratic nations are passionately addicted to generic terms
> and abstract expressions because these modes of speech enlarge
> thought and assist the operations of the mind by enabling it to
> include many objects in a small compass. A democratic writer
> will be apt to speak of *capacities* in the abstract for men of
> capacity and without specifying the objects to which their capac-
> ity is applied; he will talk about *actualities* to designate in
> one word the things passing before his eyes at the moment;
> and, in French, he will comprehend under the term *éventualités*
> whatever may happen in the universe, dating from the moment
> at which he speaks. Democratic writers are perpetually coining
> abstract words of this kind, in which they sublimate into fur-
> ther abstraction the abstract terms of the language. Moreover,
> to render their mode of speech more succinct, they personify
> the object of these abstract terms and make it act like a real
> person.

He goes on:

> These abstract terms, which abound in democratic languages,
> and which are used on every occasion without attaching them
> to any particular fact, enlarge and obscure the thoughts they
> are intended to convey; they render the mode of speech more
> succinct and the idea contained in it less clear. But with regard
> to language, democratic nations prefer obscurity to labor.
>    I do not know, indeed, whether this loose style has not
> some secret charm for those who speak and write among these
> nations. As the men who live there are frequently left to the
> efforts of their individual powers of mind, they are almost always

a prey to doubt; and as their situation in life is forever changing, they are never held fast to any of their opinions by the immobility of their fortunes. Men living in democratic countries, then, are apt to entertain unsettled ideas, and they require loose expressions to convey them. As they never know whether the idea they may express today will be appropriate to the new position they may occupy tomorrow, they naturally acquire a liking for abstract terms. An abstract term is like a box with a false bottom; you may put in it what ideas you please, and take them out again without being observed.

Anyone who wants to study the prehistory of the style now represented by "plastic words" or "connotative stereotypes" will need to take note of the *Basic Historical Concepts* and such societies as revolutionary France and Tocqueville's America. When Koselleck speaks of "the fundamental alteration in meaning of classical topoi," a steeper grade of abstraction is less decisive than a change in the nature of the concepts themselves. "The relation of the concept to what is conceptualized" is inverted; it shifts "in favor of linguistic anticipations that are meant to mold the future. In this way concepts come into being that go far beyond what can be empirically demonstrated, but without yielding up their political or social importance." When a backward look at history becomes a forward look at the still-to-be-shaped future, one is reminded of Tocqueville's false-bottomed box.

It is important not to blur the difference between the "basic historical concepts" of the epoch from 1750 to 1850 and the most recent expression of historically influential concepts which we are testing here: a peculiar naturalizing and objectifying of the historical world can be observed in this final phase. It is as though the box of abstraction has entirely lost its bottom. History has become a laboratory. The natural sciences are its model. Concepts are no longer historical in the sense that they are gained through experience, "abstracted" as one said in the eighteenth century. History is almost completely banished in these concepts. They are not concepts grounded in history, nor are they "anticipations" in Kosseleck's sense, but rather they are ahistorical manipulations of the world. Could it be, asks Jürgen Schiewe, that in these "amoebas and stereotypes" and in the "mathematization of ordinary language" a new and profound reworking of political and social semantics is evident? A new watershed? An essay such as this can only make suggestions. We will begin with "information."

When measured against what "informatio" and "information" meant from antiquity until the middle of the twentieth century, the word has been fundamentally altered. Allow me a digression:

I know an old gentleman who reads newspapers with a ballpoint pen in hand. He reads a great many newspapers and magazines and underlines at the same time; I have had in my hand articles that he had gone over and have wondered on what basis he underlines. Finally it became clear to me that he underlined according to no principle. Almost everything was important for him.

In his apartment a number of appliances operate simultaneously. When he doesn't want to miss an important radio program and something interests him on TV at the same time, he tapes the radio program. Or vice versa. He is constantly moving between rooms. He never misses the eight o'clock news.

This is no invention. The gentleman is about seventy-five and has been doing this for a long time. One could of course call him strange. I would say that he has understood something. He has understood what information is. Information is what one has always just missed.

The minister-president of Baden-Württemberg, Lothar Späth, has also comprehended this in his own way. His last book has the title: *Turning Toward the Future: The Bundesrepublik Becomes an Information Society.*

And the minister of the German postal service has understood it. He insists: "The raw material of information must be utilized." Today scientists are already calling information the fourth factor in production, beside work, land, and capital.

Every morning the "population" of the country hangs on every tidbit that calls itself "information." This ritual meal repeats itself at regular intervals during the day. It follows a liturgically regulated sequence, which simplifies an endless world that no head, no eye, and no heart can understand. The news rituals of the media articulate time and promote respect for information. Harry Pross has described this absurd phenomenon helpfully in his book *Duress: An Essay concerning Symbolic Force.*

The publishing market has understood the value of "information" for a long time. Eighty to ninety percent of its production is sold as "nonfiction."

Advertising has understood it: it dresses itself up as "information." Companies that peddle information machines and have sales of up to 50 billion dollars a year inundate country after country with their propaganda. The inhabitants of these lands, who have already been turned into the "Pepsi generation," have now become an "information generation."

Even exhibitors at a garden show in Heilbronn are informed that garden shows can offer nothing fundamentally new because the "core information" always remains the same. "Core information" in this instance refers to flowers and their arrangements.

And the young man who spoke to his very ill mother in the hospital has understood it. Afterward he did not go to see his sister in their home town to tell her about their mother: "But you phoned her yourselves; and I didn't have any new information anyway."


"Information" is a word derived from the Latin. In classical Latin "informatio" means "training, instruction, correction" or "image and imagination." The verb "informare" has the same wide spectrum of meanings. In the Latin of the Middle Ages the word acquires the additional sense of "inquiry" and "investigation" ("inquisitio").

In the Middle Ages the word also grew a German branch. Mystics had "inbildunge" [Eckhart], occasionally also "informunge" as a mirror-image reflection [reflex?] of "in-formatio." The literal meaning of the "in-building" [Einbildung] into the soul—it is God who is in this sense "built in" [eingebildet]—pales later into "engrave" [einprägen] and in early modernity becomes "to imagine" [sich einbilden]: "imagination" [Einbildung] in the present sense of a mad illusion.

The Latin word "information," or sometimes "informacion," hardly appears in German dictionaries until the early nineteenth century. Even in the later nineteenth century and on into the twentieth century it is still often left out. For example it is missing in Grimm, Paul, Weigand, Heyne, and Trübner (1943). But it can be found in Campe's dictionary of foreign words in 1801, and, beginning in the second third of the nineteenth century, in several general German dictionaries, for example those of Kaltschmidt, Sanders, and Wenig. So it has belonged to the German educational vocabulary since 1800 at least, and the breadth of its meaning has increased. Campe names only three definitions: "training, instruction, correction." Kaltschmidt adds four more: "inquiry," "investigation," "message," "report." The broadest spectrum can be found in the twenty-first edition of Heyne's dictionary (1922), where he names eight definitions: "training, instruction, correction; also court inquiry, investigation, asking after; message, report, evaluation."

This suggests that the word "information" was an abstraction that until the 1920s, had the customary characteristics of a vernacular word: it had more than one meaning, it was flexible, and it could take on various special meanings or nuances according to the context. The boundaries were not entirely fixed. The word had no universal meaning, but rather three basic particular meanings: *instruction* — here it pointed into the institutional domain of instruction; *inquiry, investigation* — here it most likely pointed to the area of jurisprudence; *message, report, evaluation* — here it probably pointed to the area of institutional assignments and messages.

In English the word had roughly the same range of meanings. It could refer either to an item of training or instruction, or to the act of molding the mind or character. It could refer to a complaint, charge, or accusation against someone. Or it could refer to news or knowledge concerning some particular matter.

From "information" as an action in time or an event, right up to the description of a result, and the use of the word as a name for an object, the transitions were smooth. The word could take each of these various meanings, according to its context; but it obviously tended more to one meaning than to another. For example one can see the progression of its variants — "explanation," "instruction," "report," "evaluation" — as a continuously graduated series between the two poles that we have named. The overwhelming emphasis was on the course of action, not the goal.

This meaning now belongs to the past. Since the 1970s dictionaries have reflected a drastic change in meaning. Its definition has been flattened and reduced in a way that seems almost reckless. The word comes to have only one meaning, in the most varied private and public spheres.

Dictionaries only begin to register the change at end of the 1970s: the spectrum of meanings has become narrower, the multiplicity of meanings has almost disappeared. "Training," "inquiry," "investigation," "instruction," and "evaluation" are not named anymore. "News" is added. The meaning has shifted completely, away from something happening in time toward its target. "Information" has become predominantly a description of a result or of a kind of object.

This change in the word's definition stems from its involvement with science. In the 1950s and 1960s the word was taken up and reworked by the sciences of cybernetics and information theory. It is now a returning emigrant: its everyday usage has undergone a scientifically authorized expansion and is supported, strengthened, and extended by a prior and parallel scientific usage. It has taken on scientific dignity.

At the same time the two definitions—the one from natural science or cybernetics and the one from the vernacular—are fundamentally distinct. In his essay "What Is Information," Bernhard Hassenstein writes: "The concept of information derived from natural science and cybernetics recognizes an 'information content' regardless of whether any particular information is right or wrong, whether it has been understood or not, and even whether in a particular case—such as with gambling—anything is even expressed." He describes the difference in this way: "In everyday language we speak about information when facts or other contents are represented verbally or by means of other devices and made available in this form to uninformed receivers. In cybernetics, however, the information content of a phenomenon, symbol, or signal has the same significance as a quantitative statement about the probability that something will appear where it is expected."

Some have wondered if it wouldn't have been wiser to avoid using a vernacular term in a scientific context. Wouldn't it have been better to create a new expression, perhaps to choose the term "negentropy"? This word describes something that has been selected or determined from a stock of possibilities.

What interests me here is that the scientific definition of information that has been in the dictionaries since the 1970s is utterly inadequate, whereas at the same time the word has had a public career in the vernacular as a *fictitious term*, a common expression with scientific pretensions. The dictionaries of course are no longer in a position to render the *scientific* term. If one examines their rendering of the *vernacular* concept more closely, for example in the Duden (1977), the Wahrig (1980), the Brockhaus-Wahrig (1981), it can be seen that the definition "findings," with its suggestion of science, dominates, whether it refers to an action or a process or, more usually, to an object or a result. That almost corresponds to the old spectrum of uses. But in uses such as "for your information," "the information is not sufficient for me," "according to unconfirmed information," the shift to a result becomes clear. The word can be recognized even more clearly as the result of a process or as an object when one looks at the verbs usually connected with it: information is *gathered, transmitted* and *distributed, collected* and *received, checked, suppressed, exchanged,* one allows it to *leak out,* one *has* and one *gives* it. These are quite colorless verbs that are used here; in the most extreme case information is merely *passed on* or *delivered.* In such cases the verb is virtually robbed of its independent meaning, it is left with only a functional grammatical significance. It merely serves to register whether the dimensions of the "information" are diminishing or increasing.

Numerous compound terms that have been introduced since the 1970s point in the same direction. The six-volume Duden dictionary of 1977 already names thirty-seven; the six-volume Brockhaus-Wahrig of 1981 lists fifty-nine combinations with "information": "Information exchange" and "information bank," "information need" and "flood of information," "information deficit" and "information gap," "information material" and "information booth"—all these are examples of compounds that reshape the word "information" into a material substance.

So an effect is projected on the word "information" just as much by the verbs connected with it as by the substantives with which it is joined. In linguistics one speaks of transfer features. When words like *exchange, need, deficit, content, flood, avalanche, gap, value, advantage* —as the so-called basic word—are repeatedly coupled with the "determinative word" *information,* then this word, determined by the returning context, takes on part of the meaning "(deficit or surplus) substance" of the basic word. And when it is mainly bound to verbs like *collect, pass on, suppress, gather, exchange, transmit,* and *rework,* then only a substantive can be put into the empty space opened in the sentence by these verbs, a substantive that has "material" properties, since it is outlined in that way by these verbs. The meaning is reshaped and secured by these new connections. But their effect extends further. In the introduction I asked whether it is possible to analyze words in isolation. I justified this question on the basis of my observation that in everyday life, people think less and less in sentences and allow themselves to be led more and more by words. This happens not only at conferences, but in public and technical discussions extending over months and years. It is not uncommon for these discussions to be determined by single words, key concepts by which whole fields of experience can be "surveyed."

This phenomenon can now be examined more closely through the example of the word "information." Such a word can take on a life of its own by establishing itself in a network of reliable supporting associations, or contexts, so that it begins to wander like the bridge supports in the 1937 picture *The Revolution of the Viaduct*, by Paul Klee. Colorless verbs enhance its effect within this static network of associations. The formation of series of combinations with objectivizing substantives raises it to a material stuff. And the frequent appearance of the word, its reappearance in countless connections—so that it seems as though at many spots in our society a dervish is throwing out the same word at intervals—allows it, so to speak, to become an oracular authority. It becomes an inspirational text. The sound alone fascinates. In a certain sense we may examine our words in isolation. The new sound "information" makes waves.

So the meaning of the word "information," once it has acquired its scientific veneer, is unified and objectified. Its meaning becomes nearly universal. And it becomes—interestingly now! after its journey through science—emphatic.

This power of emphasis is a result of the dignified aura the word gains through its association with science, and of its being used like a scientific term within colloquial speech. In science the word is used as the opposite of "redundancy," and of the excessive or unnecessary "noise" that envelops every piece of information as an inevitable consequence of its transmission. In ordinary language the concept contrasts in a similar way with excessive talk. "Information" is the hard core of meaning in an utterance; that is the sense it brings with it into the vernacular.

As a result, "information" becomes superior to mere opinion, or only intuitively grounded suspicion, or even feeling. It is fortified with data. It can be checked. As a datum, it is the essence of the thing.

It opposes everything that is not information. As soon as it is quantified, its opposite inevitably becomes a zero. One begins to feel the "information gap" or the "information deficit." Something called an "information advantage" also appears. The person who possesses information has preeminence, even in the everyday world. There is such a thing as "freedom of information" and "the need for information." It is a "good," a value. What in 1830 was called "openness" by political liberals opposed to secrecy in government now becomes a demand for access to the masses of data that remain after the scientific reduction that creates "information." It becomes a right of access to the endless and impenetrable domain of verifiable facts.

So the meaning is new; but is the word also used in a new way? What distinguishes the *plastic word* "information" from a *scientific term*, a *piece of jargon*, a *cliché*, an *empty formula*, a *slogan*, a *buzzword*, or even an *abstraction*? It might, of course, be any or all of these things. But does it also belong to a distinct class of words capable of more general definition?

We have already seen that the vernacular meaning of "information" corresponds only distantly to the *scientific* or *technical term*. But on the other hand, as we have seen, its journey through science has hidden its human face. It has been dehumanized in a way that furthers its career and allows it to gain dominance. It evokes the dignity, the luster, and the prerogatives of science; and it anchors a hierarchy and a belief in information in the vernacular.

So drawing a boundary here is easy. After all, our theme is precisely the way scientific terms change their meaning in the vernacular. In ordinary language scientific words are no longer scientific terms. In every case our examples address something different in science than they do in the everyday. In the vernacular a scientific term becomes an *amorphous plastic word*; but this change is obscured by the appearance of the word, which remains the same. By this identical appearance the word forms a bridge, linking two domains. Yet at times the meanings are so different that there ought to be two words. It is important to keep this firmly in mind, because the elites who scatter these words throughout the vernacular particularly like to speak in the name of science and enlightenment, and they very much dislike being reminded of the dialectical tension within these terms. Consider the contrast:

A scientist is fundamentally the master of his language. It is precisely his job to investigate and introduce new concepts and, if necessary, to give them new names. The word or sign that is used serves primarily to indicate something briefly and unambiguously. Such a sign should be limited in its range of applications and free of connotation. That is why scientists often prefer linguistic material that is not embedded in the sound- and meaning-fields of everyday talk. They use abbreviated symbols, proper names, and Greek or Latin words, which prejudice the concept as little as possible and allow it to stand only for its freely defined content.

The user of amorphous plastic words is much more likely to be a slave to the words. He cannot check them; instead he may have the illusion of viewing a territory in a comprehensive way. The word has a primarily social function and an "aura." Even the Greek epsilon, which stands for energy in Einstein's famous formula, can have an imposing aura in everyday speech.

There is no doubt that the transfer of scientific ideas widens the horizon of the vernacular. The history of the movement of scientific concepts into the vernacular is a history of enlightenment, and it mirrors a broadening of knowledge. But we have good reason to remember that this current has an increasingly dangerous undertow. Because a scientific term contributes to the progress of knowledge, its presence in the vernacular hints that here too a constant progress in knowledge is the norm. The plastic word becomes the means by which two kinds of progress are confused. On the donkey-bridge of the linking language, science's supposed freedom from value judgments turns inside out and becomes an even more doubtful freedom from value judgment in the social world.

Are amorphous plastic words *clichés* or *buzzwords?*

"Information" has at times been a fashionable buzzword taking its place in empty phrases. Fashion is cyclical. Something that at first is chic, highly regarded, and constantly imitated eventually grows stale and falls out of favor. In this sense the word "information" comes and goes in phrases like "just for your information" or "on a point of information." It can become an empty cliché. It is used as a freely interchangeable formula, which is useful because it is so general and yet gives the impression of saying something.

But "information" is far more than a lightweight and short-lived cliché. It is too thoroughly anchored in the structure of our world to be a lightweight. Its fashionableness is a stage from which it exerts a mesmerizing effect. The same is true when "information" appears in an empty phrase. It gives an impression of emptiness—this impression is a constant feature of any text invaded by amorphous plastic words—but its effects far exceed those of clichés and empty phrases.

Can we call these words *"catchphrases"* [*Schlagwörter*] or *"slogans"* [*Parolen*]? We encounter "information" in both guises. The concept is ascribed to entire societies and epochs. "Information" is given out as their "nature" and then becomes a program. "The Federal Republic as an Information Society." "The responsibilities of the postal service in the information age . . . " The verb is missing. In our minds we add an indicative: "The Federal Republic is *becoming* an information society." "We *find ourselves* in the information age." But what is meant is an imperative. It's an old propaganda trick to present a desired image of the future as a present reality, the hoped-for history of tomorrow as the nature of today. We are required to accept the following: In the information age a human being is a creature in need of information. The surrounding indicatives are hidden instructions, slogans conceived in the hope that they will come true.

Catchphrases *contain* instructions for action, slogans *are* instructions for action. The catchphrase "the information epoch" has a number of things in common with the words we have identified; but in what way does it differ? What, for example is the difference between "energy" and "the energy crisis"? "The energy crisis" interprets history. Catchphrases seem to organize historical fields at a glance and to reduce historical moments to a single formula. You could write the history of our century, or at least the history of what Heinrich Mann calls Germany's "public soul" by listing a series of such phrases: "the struggle for existence," "a place in the sun," "the yellow peril," "total mobilization," "the decline of the West," "total war," "zero hour," "the iron curtain," "currency reform," "reconstruction," "a planned economy versus a free market," "the economic miracle," "the crisis in education," "the energy crisis," "quality of life," "freedom or socialism," "the economic crisis," "the information age."

So catchphrases are interpretations of history. They are a static précis, fragments of sentences turned into formulas. They share attributes of our plastic words, in that they are autonomous and outside the control of the speaker. But our plastic words do not have the defining power, the picturesqueness, or the polemical pointedness of catchphrases. They are far less conspicuous and apparently more factual. They also interpret, but without aggression. They are in fact completely nonaggressive and nonpartisan. They interpret nature, not history. Expressed more precisely, they locate history within nature, and it is just that way that they achieve their dreadful effect.

Our words also differ from *slogans* like "The Saar is German" or "Uncle Sam Wants You." Whereas slogans are tied to actual history, our words do not designate a target in an expressly strategic sense. Their target is matter-of-factly contained within them.

The words "development," "progress," "information," and "education" are often used in slogans. But they also exist as neutral, scientifically based, apparently natural conceptions, made to seem self-evident not only by slogans but also by countless supporting expressions. These are words that have become commonplace concepts.

In their emptiness and diffuse generality, amorphous plastic words are not picturesque or aggressive or target-oriented, but apparently neutral. They wander in inconspicuously — bridge supports of sanctified scientific origin — and they interpret the everyday.

This raises the possibility of regarding them simply as abstractions. Can our words be distinguished from abstractions?

I think not. There is already a clue in the fact that we have not really encountered the conventional opposition between thought and feeling, naive experience and bald abstraction, with which the excerpt from Nietzsche is concerned. Amorphous plastic words are abstractions in a way that is so specific, that this characteristic only tells half the story.

An abstraction is customarily defined as an expression that evokes no sense impressions. Lifted out of the world of concrete experience and feeling, a result of intellectual work, it "generalizes" what it encounters: it withdraws from individual and historical circumstances and concentrates on what remains generally true. Not "sugar maple" but "maple tree," not "marriage" or "friendship" but "relationship," not "enlightenment" or "instruction" but "information." That is the direction in which abstraction moves. In the vernacular these steps are taken tentatively— they are handy but still only suggestive. In science they are reworked into tidy ladders by which one climbs to the level of greatest generality. And, as one climbs and the view expands, prior steps and distinguishing features disappear.

Linguistics gives us three further attributes of abstractions: (1) they have the capacity flexibly to copy varied object fields with indistinct boundaries and comprehend them; (2) they objectify sentences; and (3) they tend toward hypostasis, or personification, of the realms they comprehend.

With respect to (1): Vernacular words are vague, diffuse, and elastic, and only become precise in a particular connection. This is what makes it possible for them to indicate a heterogeneous reality both flexibly and precisely. In both scientific and vernacular spheres they are capable of transmitting precise concepts. In science they owe their precision mainly to clear, constant, and sharply outlined definitions; in the vernacular, to their flexibility. Abstractions in everyday use don't replace or suppress other ways of saying something, but rather they grasp their objects, and give them secure but pliable outlines, only on a higher level of generality. It was this attribute of the vernacular that prompted ethologist Bernhard Hassenstein's extravagant praise in his essay "How Many Seeds Make a Pile?" The vernacular, he wrote, is flexibly precise, which makes its concepts more useful than those of science, which are classically defined by their sharply drawn boundaries.

The word "information" in its older meaning was this kind of elastic, flexible abstraction; in its newer meaning it is much more a universal stereotype employed as a block, and obliterating entire fields of precise expression. That is also true for other plastic words.

With respect to (2): A second characteristic of abstractions is that they substantivize, as linguists say, the content of sentences. The abstraction "redness" implies the sentence "something is red," the word "hunt" implies "one goes hunting." "The genuine abstraction is always represented, in descriptive language [*sprachlich-descriptiv*], as the objectification of the sentence-content from the point of view of the predicate" (Walter Porzig). Often the abstraction is formed from the predicate, so that *liegen* (lay, as in the English phrase, the lay of the land) becomes *Lage* (location/position). In this way what was formerly a description becomes usable as a concept, and with the formation of a concept all the operations possible with a substantive can then be carried out. So abstractions are sentences that have materialized and now haunt the language.

In the case of plastic words they are fixed sentences, reports, and judgments about the manifold reality [Gegendstandsbereiche] of our world. As stereotypes that objectify the contents of sentences according to their predicates, they are frozen judgments. This gives us a far-ranging explanation of how they work.

With regard to (3): The third thing is that the forming of abstractions misleads us into imagining what is contained in them as a substance or a person. "Nature," "history," "language" — entire worlds of appearance become things in themselves. "Language," for example, appears as an object, and perhaps even a subject. It begins to act. Sentences like this one become possible: "So language forces us to see an independent object that has been cut loose in words like 'trip,' 'hip,' 'family,' 'wave.'" In other words language *itself* misleads us, it forces on us the tendency "to objectify, and if possible to personify, every phenomenon of every kind, insofar as it can be described by a word, and to endow it with an independent existence cut loose from other phenomena, and thereby to elevate it to a pure substance" (Ernst Leisi). This is what is meant by hypostasization.

We have pointed out that the concept "development" was hypostasized in just this way. Is it possible that the idols I call plastic words are particularly apt at objectifying themselves and then taking on the appearance of active subjects? Didn't "information" also become the subject of a history some time ago?

After this detour it is possible to grasp more clearly what we are objecting to: not to abstractions in general, but to a certain class of abstractions — connotative stereotypes drawn from science that function

as ciphers in the vernacular. They drive flexibility and vividness out of
the vernacular. Vernacular words, which can be used in a way that suits
the moment, do not drive out or replace other words. Our words, on
the other hand, dress up in the authority of science and its claim to a
universal power of explanation and then occupy the vernacular with its
highly differentiated fields of expression. They are sentences that have
become stereotypes, predicates that have petrified. These predicates,
objectified, and perhaps even personified, take on an independent existence and begin to make waves. Eventually they become the key concepts of our everyday anthropology.

So it turns out that amorphous plastic words are neither fundamental concepts of history nor scientific terms nor simply abstractions from
the vernacular. They are also clearly distinct from the darlings of language
critique—current clichés, empty formulas, catchphrases, and slogans.
They are a new word type, with whose help a new epoch is prepared
and given expression. The article by Wieland to which I referred in the
first chapter (page 19) about the alteration of the term "development"
between the eighteenth and the twentieth century is a premonitory
confirmation of our hypothesis.

We are dealing with a new type of language usage—one might call it
modular—and a new word type—plastic. In this connection, it is help-
ful to recall two analyses from the 1950s, Roland Barthes's *Mythologies*
and Günter Anders's "World as Phantom and Matrix: Philosophical
Observations about Radio and Television." Both works are germane to
my analysis here. Let me quickly sketch our areas of agreement.

"So I suffered," writes Barthes in his foreword, "from having to watch
'nature' and 'history' constantly confused with one another and I wanted
to ferret out the ideological error hidden inside the pretty presentation of
the obvious and taken-for-granted, an error which in my opinion lay hidden there." Barthes's everyday myths, which can appear on a magazine
cover or in a photography exhibition, in pinball machines, the latest
model of Citroën, or even in the new material "plastic," have a feature in
common with plastic words: they oscillate between what they report on
and what they signify. Barthes cites a famous picture from *Paris-Match*.
It is captioned: "A Black Salutes the French Flag," but it signifies: "French
Imperialism." What the photograph reports becomes almost insignificant

in the face of this meaning. The meaning sucks the pith from the private history of that saluting African and leaves it empty. In fact it denies history altogether. "The myth comes into being precisely at that moment when French imperialism passes over into the condition of nature. . . . With this we arrive at the actual principle of myth: it transforms history into nature."

Günter Anders's well-known analysis of what happens when the phantom world of the news enters a household also suggests a whole series of parallels with the way plastic words work. Both make reality unreal and both are hidden judgments—descriptions [Aussagen] that become stereotypes. "Through every item of the news the object itself is *withheld* from the audience. It remains behind in the dark, while its isolated predicate is delivered." In the news image the world is made to disappear, but exactly this fact is suppressed, since the news is presented as unmediated experience. This "ontological ambiguity" is the core of Anders's analysis. The distinction between experience and being informed of the news, between presence and absence, is extinguished for the listener. Her own four walls, and her own daily existence, cease to matter. They become phantom-like. "When what is far away moves too close, the nearby becomes distant or is distorted. When the phantom becomes real, reality becomes phantom-like."

What do these parallels signify? Barthes characterizes the everyday myths and Anders the phantom world that is inserted between the viewer and reality by the media, in concepts that illuminate important aspects of plastic words. Does this mean that the world of visual signs which concerns Barthes and Anders is beginning to have an impact on vernacular language? Did language resist colonization, at least for a while, because it still dragged so much of history along with it? Or have we just been slow in noticing?

Plastic words are points of crystallization that order the in-between world of our everyday language. The phantom world of the media and the things that have melted down to signs also resemble this in-between world. They contribute to the forging of this little set of words. But these words not only determine consciousness. As former historical concepts, now cut loose from history, they become instruments of manipulation and generate blueprints of a new reality. They are tools for the laboratory of the real.

# 3

# Plastic Words as Building Blocks of New Models of Reality

P lastic words are spotlights shone into every corner of the world. Their wanderings in science have turned them into blueprints; and as such they become the patterns of a new reality. The distance between model and reality seems to shrink as language is reduced to a series of plastic modules.

Before we enter the laboratory of the plastic words I want to give several examples. They are taken from the speeches of three West German chancellors, Konrad Adenauer, Willy Brandt, and Helmut Kohl, in the years 1953, 1973, and 1983; from news clippings from East Germany; and finally from the city plan for Freiburg, a middle-sized city in the Black Forest region.

In the state address of the German chancellor in October of 1953, there is a passage that two generations before, at the end of the nineteenth century, might easily have been taken for the work of a mad dictator:

The reason why agriculture has not been able to share in the general improvement of the last year and a half, is that at this time the price index for agricultural inputs, particularly machinery, is significantly higher than the price index for agricultural products, both calculated on the basis of 1938. This is where, in concert with industry, we must intervene. . . . A fundamental source of difficulties is the current agrarian structure. Out of a total of 14 million hectares of agricultural land, 7 million still need to be converted; that is, half of our available arable land is so splintered that mechanization in the interest of increased productivity would necessarily be unsuccessful. Numerous small and medium-sized farms are so hemmed in by the growth of villages that the application of recent agricultural methods and labor-saving machines and appliances is impossible. . . . Regulation of hydrological conditions that interfere with production is urgently necessary. . . . Rarely does the alteration of these conditions lie within the capacity of the farmer himself . . . the number of the commercial farms that suffer under these conditions can be counted in the hundreds of thousands. It is therefore necessary on purely economic and humanitarian grounds, to improve the agrarian structure. This will also bring about healthy working conditions.

The brutality by abstraction that Ernst Jünger diagnosed in "total mobilization" in 1930 is particularly evident in this example. The whole of agriculture is translated into the language of industry, and this transformation shows it in a new light. "Production," "productivity," "current agrarian structure," and "production units" are the handy tools of this transformation. The "current agrarian structure" refers to a form of farming that covered large areas of the country and had endured for centuries. As a whole the text is not yet typical in our sense; it contains relatively few plastic words. Some expressions are on their way to becoming plastic words, but because these elements of the text are premonitory, they still stand out from the rest of Adenauer's speech. The phrase "general improvement" is a precursor to the later code words "development," "progress," and "growth." "Fundamental difficulties" anticipates the plastic word "problem," and "the employment of recent economic methods" expresses indirectly what is summed up in the later favorite "modernization." These phrases—"general improvement," "problems of the present agrarian structure," and "modernization"—are the three great handles of the Adenauer mobilization. In this passage one can watch the manufacturing of a vocabulary for the most radical transformation ever to affect the farms of the rural region ruled by the Federal Republic of Germany. In Adenauer's speech the necessity for a revolution that will reinterpret agriculture as a purely economic domain in need of industrialization is treated as obvious. Farming is transformed into a domain of development.

Twenty years later, on 18 January 1973, the social-democratic chancellor of West Germany spoke mainly about "continuing solid progress" and "structural politics": "Structural politics must introduce processes of sectional and regional change, and must minimize social risk, without obstructing the necessary adaptive processes of the economy as a whole. When undertaken at the right time, structural changes also improve the conditions for political stability in the long run."

> It is becoming possible to have "a structural politics made from a single mold." At the same time as the priority of Berlin and the border regions is maintained, the enhancement of productivity in rural areas will remain a focus. Concentration of our resources on existing sites capable of development ensures the highest efficiency. At the same time the common goal of improving the regional structure of the economy demands new forms of cooperation between the central government and the regional governments. I repeat: the modernization of our agriculture is a major political task.

Willy Brandt's speech is bathed in amorphous plastic words, but it would be a mistake to regard his nebulous text as being without effect. It was the preamble to profound changes. The language of the political left has been involved with science from its origins; it argues decisively in the tradition of the Enlightenment. Brandt's speech is densely packed with words that have wandered through science.

Conservatives use this vocabulary more sparingly; but their basic assumptions are the same. Plastic words are packaged differently in their discourse, where they are cushioned by a colorful homespun style and by borrowings from religion. In the same way as the left tends to monumentalize science, the right tends to monumentalize religion. The CDU chancellor Helmut Kohl explained on 4 May 1983:

> We, the Germans, must consider our history for ourselves in all its greatness and its wretchedness, must take nothing away and add nothing. . . . The coalition of the center stands for freedom, responsibility, and humanitarianism. We accept an obligation to realize the fundamental law we have inherited from Christendom and the European Enlightenment: the free unfolding of the personality in the context of responsibility for our neighbors. This is what determines our vision. It is a vision of one people, of our people, which preserves itself by pulling together; and, by this cooperation, gains the ability to help others. . . . We will assist the countries of the Third World to develop their powers of invention and their dynamism. At the moment they are concerned with meeting their basic needs, with feeding themselves, with the exploitation of their energy resources, with education, and with the preservation of their natural environment.
>
> When we help the countries of the Third World, we are helping ourselves, for that way we also create employment in our own land. . . . I can only cordially invite all of you to participate in the great task of rent reform. In the interests of the tenant, ladies and gentlemen, we want to strive for a comprehensive solution. We will introduce a solution.

The language is a solemn and old-fashioned German, blue-eyed and upright. Yet the plastic words arise out of the ocean of pathos like islands and let us know what's at issue. There is a traditional, at times almost pathological, mistrust of anything "intellectual," of the coldness of science and the technical vocabulary, above all, of "foreign words." As impressive as this "technocratic" language once was, it is no longer popular; and this perhaps explains the worldwide victory of the conservatives. They partially conceal their identity as "experts" and declare their loyalty to the "wholesome speech of the everyday." Even the minister-president of Baden-Württemberg, Lothar Späth, who actually tends to be the scientific-technical language type, likes to speak about "visions." He blows up the fine little word "future" into a huge jellyfish with the help of ever new composites, and ends his book *Turning Toward the Future* by crowning his vision of a patchwork "reconciliation society" with a quote from the end of Friedrich Hölderlin's *Hyperion:* "Reconciliation lies in the midst of strife where all that is divided will find itself again."

So the language of the different parties is only superficially different; underneath this surface the same concepts are at work. In this respect the east and the west, the capitalist and the socialist worlds, were also identical. The selection of plastic words may have been different, as it was, for example, in East Germany, and the crystallization may have occurred in a slightly different way. The words may fit more snugly into rigid hierarchies, and the picture they comprise may have only one vanishing point; but the basic materials are very much the same.

An article from a 1973 issue of *Neues Deutschland* has the headline "Hermsdorf ceramic workers write to the ZK [Central Committee] of the SED":

> We are concerned with the further elaboration of the Marxist-Leninist class position. Our goal must be to persuade all workers through the industry-wide party organization. We must achieve clarity on all the basic questions of our time and give every worker the chance to express his creativity. We are currently focusing our efforts on fulfilling the tasks of the 1973 economic plan and on exceeding the assigned goal.... That refers to the stages of productivity in research and development, to the plan of increasing worker productivity.... We know however that further efforts are necessary in order to successfully master the task of intensification of the entire production process, by means of greater scientific-technical accomplishments, and in this way to achieve a still greater stability and continuity in carrying out the tasks assigned us by the plan.

The litany, and the interchangeable phrases, sound different, but the underlying *cantus firmus* is preserved. The word "plan" has almost a cult ring to it. In West Germany, on the other hand, the word was conceded to socialism and consigned to the devil after 1949. It was quietly replaced by the less explicit word "structure." West Germans despised the "planned economy" and instead carried out "structural politics," whether under the right-wing or the left-wing banner, in all cases using the building elements of plastic words. Viewed from the point of view of our theme the differences in the organization of these two societies are tactical. The strategy is the same, and so is the war.

Let us turn now to a small and insignificant example, the language of the city plan of Freiburg. Here we encounter the same content; the texts bristle with plastic words. The recipe of the planners seems to be approximately this: Take a handful of basic plastic building blocks, like "development," "structure," "consolidation," "center," "service delivery," "function," "factor," and "plan." Combine them with the typical technical and pseudotechnical vocabulary of the planning sector, as for example "zone," "unit," "cell," "firebreak," "axis," "density," "region," "de-coring," "traffic artery," "block," and "link." Add sparingly colorless verbs, like "documented," "contribute," "consolidate," "dismantle," "improve," "secure and strengthen," "organize," "represent," and "pursue," and season to taste with a few meaningless adjectives and adverbs for emphasis, like "important," "consequently," and "highly."

The 1978 "Blueprint for a Plan of Space Utilization" for Freiburg (whose "openness" was supposed to ensure that "as many citizens as possible are involved in the discussions about the future development of our city") can illustrate the handiness of these little building blocks (plastic words and technical words are in italics):

> Freiburg and Its Region. Goals of Regional and Rural Planning
> 1. Freiburg must attend to further *construction and development* of the spatial *structure* of the *high-density area* as a locus of particular *developmental tasks*. By means of the *consolidation* of the *service delivery* linked to a high-use *central* location Freiburg should contribute, in the framework of the completion of supervening cultural, social, and economic tasks, to the securing and

strengthening of the *functional* capacity of the *regional locus of high density.*

### The Goals of City Development

2. The *planning* goal of organizing the *step-wise* western *development of central arteries* will certainly have to be pursued for several decades longer. . . . The still-to-be-commenced *linking* of this axis to the *city ring* represents an important and consequential complement. . . . The *city-building* concept of dividing the city of Freiburg into demarcated *city sections*, each with its own *subcenters*, ought to be pursued further.

### City Construction—City Renewal

3. The detrimental tendency for people to emigrate from inner-city sectors located near the *center* can only be reduced by raising the *quality of living*. The decisive *factors* in this improvement will be gradual adjustment of the old building *structures* to the new *needs* of the inhabitants, *improvement in the quality* of open spaces (streets, squares), and *coordination of traffic* on the basis of the whole *structure* of settlement.

The generality of these excerpts is obvious. Allow me to make it even more evident by changing their context. Certain consulting companies are said to exist that present identical plans for different cities; nothing prevents us from switching the names and transplanting our texts to India:

1. Koimbatur . . . is a place facing particular developmental tasks. . . . Koimbatur should contribute . . . to securing and strengthening the functional capacity of the regional locus of high density.

Not only the location, but also the department is interchangeable. If we were to exchange the technical vocabulary of the city planning department with that of the land rationalization department in a wine-growing area, and if we chose the Markgräflerland to the south of Freiburg, we would get the following text:

1. The Markgräflerland, as a rural location with particular developmental tasks, must attend to the further construction and development of the spatial structure of the wine-growing areas. By consolidating productivity through links to a high-use central location, the wineland should contribute, in the framework of the completion of agronomical tasks, to securing and strengthening the functional capacity of the wine-growing area.

3. Decisive factors in this will be the gradual adjustment of the old structure of land use to the new needs of a sensible viniculture, improvement in the quality of the spaces suitable for grapes and coordination of traffic on the basis of the whole structure of the wine-growing rationalization region.

A final example adapts our texts to the field of health politics in India:

1. Koimbatur must attend to further building up and development of the spatial structure of health care delivery as a location of particular developmental tasks. By consolidating medical service delivery and linking it to a high-use central location, Koimbatur should contribute, in the framework of the completion of supervening health policy tasks, to securing and strengthening the functional capacity of the regional level of health.

3. Decisive factors in this will be the gradual adjustment of the old (often only rudimentary) health care structure to the new needs of the people, the improvement in the quality of the state institutions (polyclinics), and a hierarchical traffic

conception oriented to the entire health care and service delivery structure.

We could continue in the same vein. In every little sentence of synthetic words, links to other sectors can be made at will. "The development of resources," "resource information and communication." An odd rigidity and interchangeability characterizes these verbal nets. Like modules, they are simultaneously solid and easily replaceable.

Words proper to different fields can be switched around, but the inconspicuous plastic words remain. And everything else can stay too – the colorless verbs, the meaningless emphatic adjectives, and the hollow syntax. The management style resembles a rubber stamp that is interchangeably useful. We are obviously approaching the language of the experts, which I discuss in the next chapter. But at this moment we are not concerned with how a geometrically organized language is transmitted. Here I am interested in the fact that with such Lego words, models can be manufactured and these models have a mobilizing effect. The words are the constants.

Ivan Illich has pointed out to me how the usage of these words colors their environment: in their vicinity other words and word groups are taken up and redirected. The words that appear in the "space utilization plan" – "supplementation" and "accommodation" – when used in the same context as "development" and "consolidation," are barely distinguishable from them; they are semantically re-formed. The vocabulary typical for the planning sector of which I am speaking is no longer simply a technical language; often this vocabulary only gives the appearance of being technical and is actually ad hoc, made only for the purpose of the moment. In most cases words are closely determined by their context; but "land suitable for grapes" and "grape-adapted spaces" are basically nothing more, in the context illustrated above, than Adenauer's "presently existing agrarian structure." In the context of amorphous plastic words the whole semantic field begins to slide. The quotations above are just a handy, crude means of illustrating this. In this work I can merely suggest the outlines and direct attention to only the most obvious features of this phenomenon.

The initial abstractness of even a text like the "Blueprint for a Plan of Space Utilization" must not be allowed to fool us; it is more than merely hollow. It is a grader that flattens the landscape. In the train of such a shadowy machine the most astonishing projects can be accepted today, even in little villages. But the 1978 Freiburg Plan also becomes more concrete:

> For the duration of the space use plan (16 years), measures are to be expected (a traffic plan, "green" sanitation, modernization) which will require improvements to the existing building stock.
> ... Particular measures will deal with the replacement of building stock, the decoring [Entkernung] of areas encompassing certain blocks, and the improvement of open spaces.
> ... On a purely numerical basis the city is undersupplied with parks and green spaces.

Subsequently, the governor of the region of Baden-Württemberg, as was mentioned in Chapter 1, let it be known in Freiburg that the city lay in the shadow of the border and needed a major "structural leap." One can see now how this arrogant project of transforming "the sleeping city" into the "metropolis of a new area of activity" has been energetically "taken in hand." In this the mayor uses the word "citizen" so frequently that it begins to act as a kind of spell. An ambitious advertising campaign presented at a state garden exhibition, or "scenarios" unveiled at city hall, are careful to present a shining vision of the city's future: the "citizen," the subject into whose head the new city model must be stuffed, is reformulated into a creature with new "needs," to which "development," for example the "building structures," must then be matched. In this way the flesh-and-blood inhabitants of a place are turned into interchangeable units.

The language of the Freiburg Plan reveals how a historical place is made flexible and organized in a new way; how a transformable "substance" is made of what exists, and how the contours of a new crystallization of this substance are then drawn, at first abstractly and then ever more concretely.

The language sets the scene for actions that can so transform the city—there are plenty of examples of this!—that within one generation it becomes unrecognizable.

In this language there is something that, obviously or latently, has come loose: it seems to be unhinged. At the moment this loosening is still masked by the technical sound, but it can appear at any time: "For example in the west of Freiburg," writes the *Stadtkurier*, "the director of the new suburban plan draws connections like rays, with cross-connections repeated again and again. Green zones and farm areas not only form a belt around the city, but also enclose the small settlements. The planning director does not create new corridors but intends to preserve and to use the existing structures so that smaller settlements are only a short distance from their own leisure areas. . . . " "Planning in and with the existing substance," is the new motto. One could even imagine a complete transformation of the entire city. And indeed, in other places the construction fever that has swept the Federal Republic has already resulted in the demolition of a lot of buildings that had earlier managed to evade Hitler's greed for destruction. Visibility! "We must start from the absolute imperative that the city be aired out by corridors." Every little section must have its leisure area. On the one hand, broad corridors with unobstructed views must be cut through the city, and on the other, colorful structures must define the housing areas. Emphasize certain specific landmarks in the various open spaces and provide big-city nodes with large projects. The four-lane interchange and the high rises that are built around the train station will probably be followed by urban renewal of the adjacent areas. . . .

These planning games turn cities into construction sites, and in the rural areas they lead to "earth moving." In the laboratory of the plastic words, as we saw, models are developed with the simplest materials, and reality is forced to conform. It is obvious that economic interests are in control of what is done; but what is more important is that these actions take on the appearance of events, of processes, and are experienced that way: as a general process of nature that follows its own unerring path and, as Hegel says, freely steps on a number of innocent flowers. The outrage of the individual no longer seems even to touch this process; its ghastliness is barely noticed. It has become a matter of unremarked common sense. In the nineteenth century, had the state tried to transform the homeland of the farmers of the Kaiserstuhl into a synthetic, terraced landscape for wine production, the farmers would presumably have fetched their guns out of the closet and opened fire. In the last few decades they have acquiesced almost without resistance. The volcanic soil of the Kaiserstuhl, which has been under cultivation since Roman times, became a factor of production and was designated as "wine-suitable"—or in certain places "not wine-suitable." In Freiburg, likewise, there have been only a few individuals or anarchic groups that have tried to put a wrench in the spokes of this "weltgeist"—and who only manage to provide an excuse for putting "development" in place with military force, to general approbation. A historical place is reshaped into a laboratory.

Let us enter the interior of the language laboratory! Here it looks almost like a workshop in experimental and concrete poetry: but much hangs on that "almost," of course.

An alphabetical ordering of the plastic words already makes it evident that they tend to form sentences, even without verbs (which after all, as I have already said, play only a minor role):

"accomplishment"
"basic needs"
"capitalization," "care," "center," "communication," "consumption," "contact"
"decision," "development"
"education," "energy," "exchange"
"factor," "formation," "function," "future"
"growth"
"health"
"identity," "information"
"living standard"
"management," "modernization," "model"
"partner," "planning," "problem," "process," "production," "productivity," "progress," "project"
"quality"
"raw material," "relationship," "resource," "role"
"service," "sexuality," "solution," "strategy," "structure," "substance," "system"

"value"
"work," "workplace"

Some words arrange themselves into rows as if on their own. "Problem-solving strategy," for instance, is a string of plastic words that has the status of an "almost-sentence." As George Orwell and Hannah Arendt have noticed, when this metaphorical string of words is used to speak of politics, a soluble problem is implied, and when the strategy intended for its solution is war, these words become deadly. One can see, therefore, how an almost-sentence such as "problem-solving strategy" already is a fully stocked arsenal.

"System," "structure." The first word permits us to encompass a global domain and suggests its completeness—the "health care system," "our economic system"—the second suggests the architecture of an opaque shape. Even when one has neither an overview of nor any insight into a subject, one can still speak quite easily of "system" and "structure."

Presumably the highly abstract character of plastic words is their most effective property; the abstractness levels the language field and the field of the affected objects.

"Energy–resource–production–care–consumption." The chain is highly suitable for universal interventions and programs. Here the talk concerns the economy. Similar chains can be made for rural planning, education, health, or transportation. They negate all local idiosyncracy. In five words they enclose and organize enormous spaces.

"Information," "communication." We are approaching the inner circle, the words in the center of the cosmic big wheel. Here the meanings become completely universal and ahistorical and allow themselves to be mathematically expressed—and soon enter our everyday consciousness.

"Role," "factor." What is there that can't become a "factor" or play a "role"?

The words are interchangeable in a disconcerting way. They can be made equivalent and strung together like modules into a chain of equivalent sentences. Again and again they appear to make sense: "Information is communication. Communication is exchange. Exchange is a relationship. A relationship is a process. Process means development. Development is a basic need. Basic needs are resources. Resources are a problem. Problems require service delivery. Service delivery systems are role systems. Role systems are partnership systems. Partnership systems require communication. Communication is a kind of energy exchange." Some equivalents can also be expressed without a verb: "*Resource:* energy and health, information and communication. *Model:* progress and partnership."

The mobility of these words, their capacity to be linked up, is unbelievable. A friend from Marburg tried them out in the child's game in which children sit in a circle, and one child after another picks up a folded piece of paper with a word on it, and then everyone has to say their word without knowing what word their neighbor chose. The chance of hitting on a sentence obviously becomes very high when the game is limited to our plastic words; they almost always seem to make sense.

With these generalities we are moving toward a height, or a depth, in which the building blocks come to resemble one another and the possibilities of their uses are limitless.

It is like one of the nightmares of the early eighteenth century. In Jonathan Swift's *Gulliver's Travels,* Gulliver comes to Lagado, the capital city of the island of Balnibarbi, and there he inspects the Great Academy of Lagado. He is shown the projects of various professors, spread out over no fewer than five hundred rooms. One is working on a machine that will on one far-off day make it possible to write books in a totally mechanical way — without costing a lot and without the slightest need of talent or education:

> He then led me to the frame, about the sides whereof all his pupils stood in ranks. It was twenty foot square, placed in the middle of the room. The superficies was composed of several bits of wood, about the bigness of a die, but some larger than others. They were all linked together by slender wires. These bits of wood were covered on every square with paper pasted on them, and on these papers were written all the words of their language, in their several moods, tenses, and declensions, but without any order. The professor then desired me to observe, for he was going to set his engine at work. The pupils at his

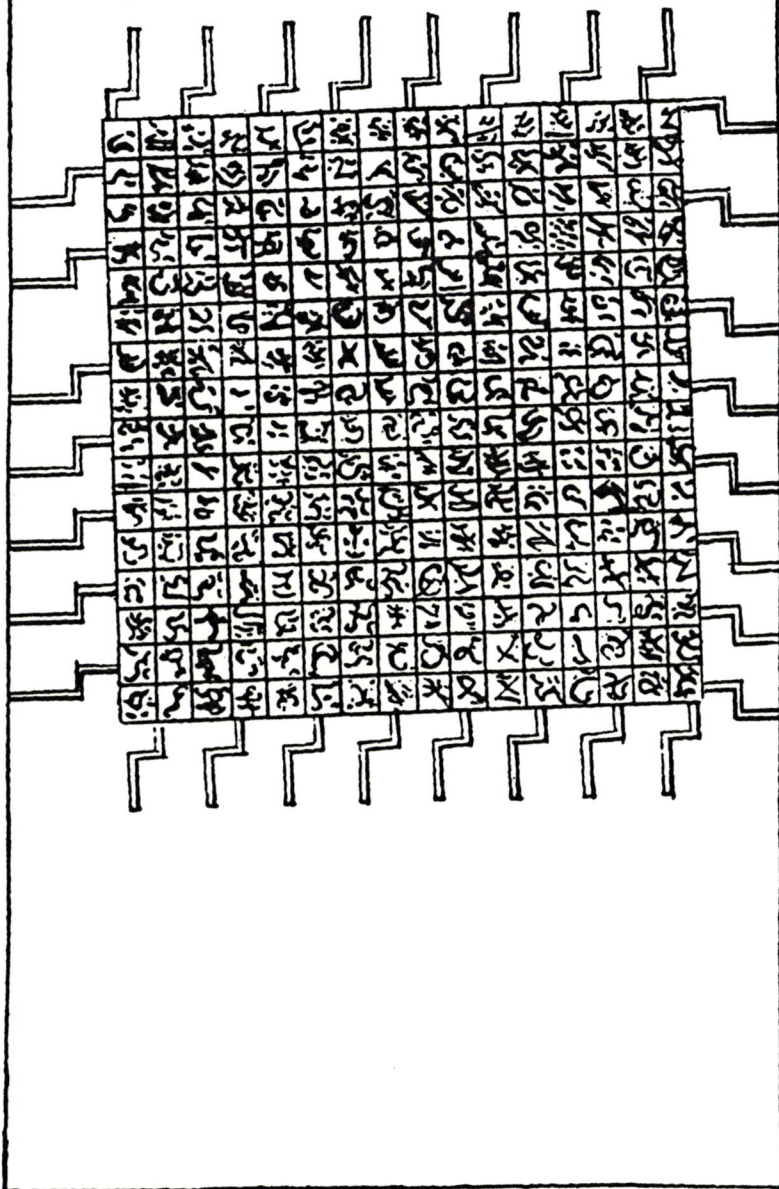

Fig. 1. The word frame of the Great Academy of Lagado. From *Gulliver's Travels* by Jonathan Swift.

command took each of them hold of an iron handle, whereof there were forty fixed round the edges of the frame, and giving them a sudden turn, the whole disposition of the words was entirely changed. He then commanded six and thirty of the lads to read the several lines softly as they appeared upon the frame; and where they found three or four words together that might make part of a sentence, they dictated to the four remaining boys who were scribes. This work was repeated three or four times. . . .

The professor said that out of these materials he "intended to piece together a complete body of all arts and sciences," and added that this could be done even faster if the public could subsidize five hundred such apparatuses, and then have the results coordinated.

Gulliver's academic utopia has in the meantime become reality. Since I have been talking to people about this theme, I have often been given amusing games. The Straelen Institute of Translation, an internationally-known center for translators in Northern Germany, has come up with a "phrase-threshing machine." In an elongated envelope three rotating cardboard disks are attached side by side, almost completely covered and only visible through three windows which are arranged horizontally in one row. If one turns the disks, three words appear at the windows side by side: in the first there is an adjective, in the second a substantive with a hyphen, and in the third the substantive that is linked to it. On the side labeled "conservative" there are then rows like "oriental culture-responsibility," "this existence-enlightenment of ours," "unshakable future-remembrance," "profound education-religiosity." The words, which are arranged on this side as though in a pantry, mostly belong to the "jargon of authenticity" described by Adorno, which circulated in the 1950s and now is being rebrewed like a tea made with used leaves. On the opposite side of the envelope, labeled "progressive," there are word sequences like these: "ambivalent fluctuation-phase," "integrated organizational-structure," "functional communications-conception," "systematic identi-fication-problematic," "permanent innovation-flexibility." So on this side there is the vocabulary of the 1970s, rather unfashionable at the moment. It is worth pointing out that the words we have been looking at—"future,"

**Phrasen-Dreschmaschine** konservativ

mannhafte Kultur- Bewältigung

aus der Wortspiel-hölle des Übersetzer-Kollegiums Straelen

Falls Sie an Formulierungsnot leiden: hier sind insgesamt 8000 Phrasen. Auf jeder Seite 1000, und wenn Sie die Bestandteile von beiden Seiten kombinieren, noch einmal 6000. Außerdem können Sie in die Zwischenräume der Scheiben Ihre eigenen Lieblingsphrasen eintragen. Das dürfte dann für Ihre Zukunft reichen.

© Klaus Birkenhauer / Straelener Manuskripte Verlags-GmbH, Postfach 1224, D-4172 Straelen 1
ISBN 3-89107-000-4

**Phrasen-Dreschmaschine** progressiv

systematisierte Organisations- Phase

aus der Wortspiel-hölle des Übersetzer-Kollegiums Straelen

Falls Sie an Formulierungsnot leiden: hier sind insgesamt 8000 Phrasen. Auf jeder Seite 1000, und wenn Sie die Bestandteile von beiden Seiten kombinieren, noch einmal 6000. Außerdem können Sie in die Zwischenräume der Scheiben Ihre eigenen Lieblingsphrasen eintragen. Das dürfte dann für Ihre Zukunft reichen.

© Klaus Birkenhauer / Straelener Manuskripte Verlags-GmbH, Postfach 1224, D-4172 Straelen 1
ISBN 3-89107-000-4

Fig. 2. The "phrase-threshing machine" created by the Straelen Institute of Translation. © Klaus Birkenhauer "education," "communication," "structure"—are represented on both sides in impressive numbers. The machine, described by its inventors as a "purgatory of word games," permits, so they claim, the production of one thousand phrases on each side, and if one combines the two sides, the number goes up to eight thousand. They use, as their starting point, the generative grammar of Chomsky, according to which a definite list of signs and rules of combination will produce a calculable number of sentences.

From someone at the Ministry of Health in the northern German state of Schleswig-Holstein I received an "automatic fast-formulating-system," which had apparently been used in the 1970s as a predecessor of the "phrase-threshing machine." On a board there are three columns of adjectives and two substantives, side by side and numbered from 0 to 9. The substantives are of the "progressive" type. One must pick a number combination. For the number 5.7.8., for example, one gets "functional growth-problematic." The vocabulary corresponds largely to that of the "machine" I described previously. The system is attributed to Philipp Broughton, an employee of the U.S. Department of Health, who apparently designed it for bureaucrats who wanted to lend their documents an air of authority. I received an English variant of this system in December 1986 from the United Nations University in Tokyo. It was conceived "to foster creativity in United Nations documents," and contains an even greater number of plastic words than do the games described up to now: 0.0.0 global management systems; 4.8.11. national education plan; 2.2.2. regional information center, and so on. Here again the combinations always make a kind of sense.

These merry games represent the condition of language; they point to a kind of loosening of our language that is historically unprecedented. But at the same time they also reflect the fact that reality itself has taken on an unparalleled flexibility. Nor do these word games only produce empty phrases or technical jabbering. That would be harmless. Their most effective products—and if one narrows the game to plastic words one can recognize this—are the finished building blocks of our world. *Natura fictionem sequitur.* Nature follows art, and at this moment pretty bad art, trash. A language of interchangeable words is reflected in a world of interchangeable trash. The principle of *grand arbitrariness* with which postmodernism pleases itself is currently leading to a trashing of history. The little cylinder-shaped projections on Legos that turned the loose wooden blocks with which children used to play into plastic modules that can be combined, have brought about this result. The utopian project of the Lagado Academy has turned into the Lego of plastic words. A modicum of effort on my part produced the following satirical collage of modular words:

> Communication is a basic need. It is a significant factor in maintaining public health in a democratic society. Ultimately, it not only fosters the exchange of information but also increases consumption and therefore, indirectly, production. It is a new and extreme form of growth. It creates value. What is meant here is communication in the broadest sense, not only artistic, political, or scientific communication, but also the so-called spontaneous process of exchange that can occur at staff retreats. —Partnership is also a many-faceted module. As a partner we each have a role and a function and become a calculable factor. We reveal ourselves as transactional partners. In the case of ongoing projects we must also unquestionably develop our capacities as discussion partners. Partners are—by definition— interchangeable. Clients count as partners. When problems arise between partners, they seek the services of experts. Partnership is a property and a process. A resource.

Is "partner" perhaps already a grammatical category in its own right? And "partner" is certainly not the only such word. As another example, a number of our building blocks can be connected with the plastic word "problem"; from the "problem of work" to the "problem of contact" to the "problem of care" this word publicly represents the idea that a need exists that must be satisfied. Composites of this type, which are recognizable because the second part makes the first more specific, are equivalent to sentences, which they resemble. Their first part corresponds to the subject, their second to the predicate. So in ever more numerous connections "problem" becomes a report, a judgment, and a call to action.

"Development" and "process" are analogously suited to the forming of series. We have production development, energy development, communication development, consumption development, and health care development. But at the same time we could also put "process" in the place of the predicate, or "system" or "structure."

In these series of linked rows, the concepts are on the way to becoming pale hangers-on, grammatical categories, suffixes. That means they already have become such commonsense ingredients of our everyday consciousness that they are hardly noticeable. They have wandered over into the elemental domain where for example the phases of time: the past, the present, the future—or of number: the singular or the plural—belong. They form our world in similar ways. This world is deficient, mobile, and resolves itself into ever new structures: that is what the elementary grammar of plastic words conveys.

# 4

## Experts as Functionaries Who Make Reality

He ere I want to examine the questions raised in the previous chapter from another point of view.

How do the words "development" and "health," "education" and "information" become texts? How is the net of plastic words woven so that it overwhelms and captures our consciousness? Who turns these basic building blocks into models, and those models into projects that bind society? Who is the agent of this colonization?

In our presentation of how models of reality take shape in the laboratory of plastic words, the languages of experts became visible. Experts adapt these elementary building blocks to the areas in which they will be used and attach them to the typical vocabulary of this area. They are the functionaries who make reality.

The expert is an authority. In German this is not the same thing as a specialist. A specialist works within his sphere, as a crystallographer, an economist, or an administrator. The word "expert" wanders back and

forth between specialized languages and the vernacular. It points backward into technical fields and forward into the workaday world. It
twitches and oscillates. Its meaning circles in the way described by
Roland Barthes in *Mythologies*. The point of view alternates between the
technical sense of the word and its nimbus. The nimbus predominates.
The word "expert" points to the authority of someone familiar with
scientific and technical matters. Experts speak with authority.

It is not only the word but also the person it describes that wanders
back and forth between specialized fields and society at large. In a telling
formulation in the most recent Brockhaus dictionary an expert is defined
as a person who "is increasingly given the task, on the basis of methodically gained specialized knowledge, of formulating goals, analyzing
problems, and suggesting measures for their control. This is true for
many areas of modern social life." The expert is a mediator. He is a
phenomenon of a world that has become mobile, a world that doesn't
have time for a slow and careful dissemination of technological advances.
He is the point at which knowledge is put into practice. He has stepped
into the place of mandarins and priests. He guarantees the progress of
civilization. Between technical language and the vernacular the expert
balances, using both sides to steady himself, and advancing toward the
general development he has projected as his goal.

What are his qualities? Can a profile be sketched of this type?

For some time now a job called "headhunter" has existed. "Head-
hunters" are employed by large firms to find the right person for a
particular position. From a certain salary level up, positions are not filled
by advertisement and application. Instead, a headhunter is commissioned
to create a profile of the position to be filled and then to be on the
lookout for such a person.

Here we are not on the lookout for particular persons, but rather for
experts in general, for a profile of the dominant type, or more modestly,
his language. Does he use a characteristic set of words? Before I undertake this sketch, I want to reflect further on the present condition of
ordinary language.

As I have already said, both our public and our private use of language
have thoroughly changed. A typical headline like "Aggression: Why

Even Small Children Bite and Hit" on the cover page of a magazine would not have been possible twenty-five years ago. Nor would people have accepted the assertions of the accompanying article: "1. Children can be made aggressive in diverse ways and 2. The strength of aggressive tendencies varies greatly." "Aggression" didn't exist as a term of public discussion, and people felt no need of expert guidance in living their daily lives. Countless other examples would be possible.

The degree to which scientific words have entered the half-private realm of ordinary intercourse is even more remarkable. A kindergarten teacher can speak of our child's "attempts at making contact" and his "social inadequacies." A clergyman who has been trained in behavioral psychology may adopt a personal tone, but he does so in a way that is so shaped by his training that his amiability is only counterfeit.

Even private conversation is disfigured, and over an even larger area than the radius of "sexuality," by a language created for other purposes.

A crust of science and technology has hardened over our common language and given it an authoritarian ring. But this language only seems technical. It is made to appear so by the ad hoc use of composite terms, by the use of neutral, objective, and impersonal expressions; and by the addition of abbreviations and numbers. This practice colors the language of politics, of newspapers, and of public discussion. New words are created daily to produce an impression of specialization. "*Metal fatigue* caused the collapse of the meeting hall." There are not only disposable diapers, but also disposable words.

The vernacular is speckled, *gestrifelt,* as one could say in the late Middle Ages when referring to a language mixed with Latin or French. The vernacular has been infiltrated by a hybrid lingo made by marrying plastic words to technical terms.

The reason? It is hardly possible to speak about this without using religious categories. The twentieth century doesn't lack faith. Science is its national church; experts administer its miracle cures. We are experiencing a restriction of conversation and a general atrophy of memory. A new, more homogeneous sense of time is evident. Things which formerly belonged to the domain of local and orally transmitted bodies of knowledge are now incorporated into a universal written culture. The alphabetization of knowledge stretches without interruption from infancy to the last rites. This emphasis on written language disempowers people by bringing material existence under the purview of experts who are oriented to writing and use writing to orient others. Knowledge from above, readily available and pregnant with authority, reaches into all local and private spheres.

Let's take as an example the apparatus of health care. No one who is healthy talks about her health. Nothing is bothering her; she doesn't lack anything. There is no reason for her to speak of this "nothing," since she doesn't notice it. She only begins to speak of it when her body forces itself on her attention: then she talks about her illnesses, if they come, or her memory of her pains. The word "health" comes up infrequently in the old texts; and when it does, it designates an absence: it means "uninjured," "alive." Whoever was healthy lacked nothing. But in the time in which we live health has become a virtue, of which we keenly feel the lack. This lack has now been implanted in everyday consciousness. So we are constantly talking about our illnesses.

When the concept of health gets loose in the vernacular, it generates new forms of deviance. Originally, it was a rather unobtrusive idea, but that was before it was authorized and sanctified by experts. Now it introduces arbitrary boundaries into the continuum of experience, erecting a barrier between "healthy" and "sick," and specifying a norm that has been set ever higher, so that ever more people are identified as sick. The new norm quickly takes on the appearance of being natural; its origin in the past is forgotten.

The vernacular now contains words like "iron level" and "blood pressure," "pain threshold" and "central nervous system," "intestinal flora" and "viral infection," "inflammation of the upper respiratory passages" and "antibiotics," "side effects" and "compromised breathing," "heart failure" and "kidney failure," "potassium deficiency" and "reduced resistance," and "chronic illness" and "acute illness." These expressions were exported into the vernacular by the health sciences, the health economy, and the health administration. They were transformed when they changed spheres. Technical meaning yielded to social meaning. The scientists' awareness that "healthy" and "normal" are distinctions located on a shifting scale was lost. The notion of a norm that emerges from this particular vocabulary, of a healthy middle range that lies between "too high" and "too low," has become a fixed standard in ordinary life. This norm is set at a level that we somehow always fail to reach. And so, because of the authority residing in a normative language, the continuum of experience is measured against a fixed yardstick, with the result that we are constantly asking ourselves: aren't we coming down with something? Isn't something wrong with us? When the yardstick is passed into our own hands in the name of prevention or personal responsibility, we become in effect our own clients. It turns out that no one is "healthy" any more.

So scientific words form countless bridges into the vernacular. But it is not the altered meaning they take on there which explains the exalted status of the experts. It is the fact that these opaque vocabularies above all shout "science!" They transmit the reputation of science into the everyday, and through this connection they awaken trust.

Max Weber remarked that most modern people don't actually know how a streetcar works, but they live in the belief that they could find out at any time: "The increasing intellectualization and rationalization therefore does *not* signify an increasing knowledge about the factors governing one's life. Rather it means something else: knowledge of or belief in the notion that *if one really wanted* one could find out at any time—that therefore there are fundamentally no hidden incalculable powers at play, that rather all things (in principle) can be *controlled through calculation*. But that signifies the disenchantment of the world." This is precisely the effect of the words borrowed from science. Countless equivalents of Weber's streetcars drive across even a single page of the daily paper. But perhaps we would no longer use the expression "disenchantment." Science not only disenchants, it also enchants. To quote Ivan Illich: "In its practical application science enlightens, replacing the fantasies that make sense of things with opaque assurances. It encloses even reality in the domain of the experts." This situation has been brought about, at least in part, by the experts themselves, and it decisively establishes their dominance.

I remarked at the outset that plastic words, insofar as they connect the world of science and the world of the everyday, function as metaphors. I would now like to explore this concept further, in order to clarify how science and technical terminology work in the vernacular.

Metaphors are figures of speech that pull two spheres together briefly. "The blade of her joy broke in two," says the medieval poet Wolfram von Eschenbach about Herzeloyde, when her husband, Parsifal's father, is killed. The proper spheres of the two terms, blade and joy, can still be distinguished. They are not explicitly compared but rather welded into a single form. The result is a tension, an abbreviated comparison. As long as it is still new and surprising, the figure can be recognized for what it is. Many metaphors from the vernacular, for example, "opposition" or "electrical flow" or "development," have become so conventional that the image they contain has faded and they no longer look like figures of speech at all. But in most cases the imagery can be recalled to memory.

It is different when the term in question derives its significance from the domain of scientific abstraction and represents nothing concrete. When someone speaks about the "social dysfunction" of his child in a private conversation, there is still a linking of spheres and therefore a metaphor, but the chasm between the sphere of origin and the sphere of application is easily overlooked. There is no tension; no spark jumps between the two spheres. They are tied together seamlessly. Their original separation is hardly remembered. The result: one takes the word for the thing.

The transmission of words and images from one milieu into another casts things in a different light. It can unlock meaning and promote recognition, it can alter perspective, and it can organize and illuminate the new milieu. The linguist Charles Bally spoke about "effets par évocation d'un milieu." But at the same time this effect is not without problems. The transmission can also disfigure the sphere of application. And not only can an individual metaphor be off-kilter and distort its object; an entire frame of reference can also be carried into a new domain and have the effect of simplifying, reducing, and alienating it. The forms of expression of one milieu can overwhelm, cover up, and "colonize" those of another milieu. There seem to be three particularly effective spheres of origin, three big image donors, which export their vocabulary and shed a strange new light on everyday existence: the languages of science (and technology), of economics, and of administration. These three "colonizers" intermingle and conquer society; they share a specialized form of expression.

Development, for example, was discovered to be the law of life and the animating principle of history in the eighteenth century; and, as such, it became a key concept of science.

In the hands of the economist this law becomes a resource. It defines whole sections of the planet as being in need of development and recognizes development as an unending project for economists.

The bureaucrat administers this project. Wherever a need for development is recognized, he takes over with his handy language. He can, for example, use the term "project" as a module that readily can be attached either at the beginning or the ending of other words ("project initiation," "project leader," "project materials," "animal breeding project").

It is only in the last few decades that "information" has become a universal key to science, and the world has been recast as an information system. But the economist has been illuminating the field described that way for a long time, for instance when he encourages us to make use of "the raw material of information" and labels it as "fundamentally an expanding market." He *makes* it into an expanding market, insofar as he is able to enforce his insight that our epoch is "the age of information."

The bureaucrat administers the information. A sizable bureaucratic apparatus is at his disposal for this purpose in the postal service; or he opens up a market for the economist by going through the school administration, which is very cooperative. There he finds a ready-made language. ("Basic computer skills," "computer-supported learning," "computer-supported structuring.")

These scientific master keys interpret the world "as it exists at this moment" in a new way and offer a new unifying order of things. The economist converts this order into an economic imperative, and it then becomes a simple matter for the administrative specialist to introduce his organizational language into these fields and to develop them—first of all in their language—into service bureaucracies. Here we are referring only to the linguistic aspect of this event, but I think this can be profitably examined in isolation. The transmission of words across boundaries is the most noticeable feature of our current use of language. Images originating in science, economics, and administration reshape even the most minute regions of daily living and colonize them.

It seems particularly apt to speak about metaphors here because languages once particular to a certain milieu have expanded far beyond the circle in which they were originally used and penetrated regions formerly foreign to them. The critique made of Karl Korn's *Language of the World of Bureaucracy* — that administrative language is a genre that has always existed — doesn't fit. He was concerned with its bursting the boundaries of its original usage. For example, once education was declared a major resource in West Germany, the number of workers engaged in manufacturing its specialized knowledge (educators of education) rose dramatically. The scientific literature of pedagogy multiplied enormously and took on the unmistakable characteristics of a mass commodity. And, strangely, its language can often no longer be distinguished from the language of administration.

Transmission of words between spheres is barely noticed anymore and, for that reason, it seems all the more commonsense. The success of the practical colonization of our world partly depends on a prior metaphorical colonization. A toolbox of modular stereotypes is available and ready to use. But this transmission not only opens up new regions, as I have said, it also disfigures and estranges them. The metaphorical colonization means, linguistically as well as concretely, a perversion of the social world.

The three colonizers do not affect the world independently. They interpenetrate one another, and their interpenetration can be very deep. How the hierarchy of science, economics, and administration is arranged may be as significant a question for the present as were Jacob Burckhardt's famous reflections on the three powers of religion, the state, and culture in history. It is easier to make out how the three spheres are bound together: specialized language is their connective tissue. Experts are their organs.

So again I ask: What are the properties of this type, the expert, who according to the *Brockhaus Dictionary* "by virtue of methodically gained specialized knowledge, increasingly has the task of setting goals, analyzing problems, and recommending measures for their control"?

I have studied countless texts, texts on city planning and the rationalization of agriculture, the scenario projected by an "Institute for Structuring the Future," the report of a commission from Baden-Württemberg called "Perspectives on Future Social Development," and Lothar Späth's *Turning toward the Future*. I've read government speeches and essays in newspapers and journals (since 1983) in which the "transformation" of the Federal Republic of Germany into an "information society" was promoted. I've collected newspaper clippings about the most varied areas slated for development and looked at advertising texts and brochures on development aid. I visited a symposium on the development of the Freiburg region and an information session with the motto "So That Our City Remains Our City"; and then I listened to the experts after the "disturbances" in Chernobyl and Basel. During this time I focused my attention on the language of the experts and composed the following ideal picture:

The expert attributes human decisions to the force of circumstances— *nolens volens*. His charm consists in seeming to assume only modest partial responsibility, and yet at the same time he lets it be known that basically he alone is responsible. Every question is ultimately a question for experts. He puts forward as self-evident and unquestionable the judgments contained in the plastic words. In his eyes, the project of recasting history as nature which these words embody has already succeeded; the historical domain has become "valueless" for him and indistinguishable from science, where this project originated. But there is more: history seems to have come to an end for him. The historical record shows that humanity has already forgotten its techniques— exceptionally fine and impressive techniques of architecture and vase painting and glass art—a number of times in the course of the three thousand years we can recall. The expert assumes that our current techniques are of permanent duration. Otherwise it would be incomprehensible that he could justify the oppression of the earth by poisonous radiation, saying it is a "fully responsible act," even though it can only be dealt with if the science and technology of today are maintained for twenty thousand years. The expert, assuming a permanent modernity, defines what is possible and indicates the boundaries of the possible; if he still asks anything at all, he asks whether something will gain general acceptance. The analyses he produces of situations, the perspectives he develops, the help he offers as a consultant, and the practical actions he recommends serve to make concrete, as representations of value, his silent certainties.

Just as he has been undermining the independence of jurisprudence for a long time, the expert is destroying the independence of moral and political decision-making. Our natural sciences have an ethos that cannot be distinguished from the absence of responsibility. *They* are his model. The domain of planning demonstrates that the expert has become the determining factor and that he "functions" in the way described here. He offers extensive data as the necessary basis for political decisions without noticing that he is only confirming the power of the categories he has created. The data and the needs calculated from them only continue the current condition and its inborn "trend." They replace politics and dissolve democracy. Do we need atomic energy? A highway through the Black Forest? A better traffic connection with the industrial north? A war in the Persian Gulf? To the degree that the answers to such questions are made to appear as the forceful, inevitable dictation of circumstance, it makes sense that decisions are increasingly entrusted to computers. Politicians no longer monitor the data that is entered and the consequent processes of calculation. Decisions are reduced to trendlines determined by data.

The expert understands something of his field; but he derives his strongest authority from the *language* belonging to it, from the stance of the nonpartisan scientific specialist. He throws the mantle of scientific language over himself and disappears beneath it; he gains distance from the layperson and through this distance wins effectiveness. Three characteristics of scientific speech serve his purposes:

— the *prestige* that has historically belonged to the language of science as the successful instrument of the Enlightenment. He borrows it for himself and always speaks in the name of the Enlightenment;

— the increasing *opacity* of specialized vocabularies. This strengthens his position, creates a contrast between a responsible elite and an irresponsible laity, and divides the world into experts with something to say and others who may hardly speak. His incomprehensibility alone already turns him into an essential institution;

— the capacity of scientific speech for *imitation*. The more fixed the language is, the clearer it is that specialized terminology, passive sentences, neutral word choices, and tables are markers for science, the easier it becomes to replicate things. It is then no longer difficult to use this loom to manufacture linguistic garments. Hans Christian Andersen's story "The Emperor's New Clothes" is the story of our time.

But the expert is also a translator. He translates science for the layperson, and not only in the sense of making it easier to understand. He is in every sense its transmitter. For he transmits that which has been newly accomplished in the world of knowledge into the social world, without acknowledging the yawning chasm that exists between them.

The expert of whom we speak does this as the representative of progress. From his position in the great stream of progress he describes everything in his domain that still hangs on from former times as out-of-date and anachronistic. The simplest way of claiming superiority without spiritual cost is to present himself as progressive. He unfurls the future like a banner. A study is not a study, but rather a pilot study; a project is not a project, but a pilot project; each modest beginning is located within a far-ranging program. The future is the space in which all his endeavors pay off.

His most convincing argument is name-calling. Anyone who refuses to get involved is hopelessly backward. Such a one is sleeping through development. The expert replaces the pair "good" and "bad" with the pair "progressive" and "backward" and in this way instills *his* order of values. Here he has a rich vocabulary available to him. On the one side the electrifying suggestion of the "modern," "the current," "the coming thing"—on the other the feeble appearance of the "old-fashioned," "the anachronistic," the "out-of-date," the "ancient." In fact it is quite possible that what is progressive in one domain will be a decided step backward in another. But he blithely transmits the basic metaphor of the sciences—their conception of a river of progressive discoveries—unaltered into society, as though change were always progress. He erases awareness that two fundamentally different spheres—the sphere of the unlimited expansion of theoretical and technical capacities and the limited sphere of daily life—are pinned together with a metaphor. This metaphor actually belongs to the first domain only and disfigures the second. In the eyes of the expert, whatever is scientifically and technically possible is also socially possible: for example, dealing with animals as patentable artifacts. Anything that is revealed and discovered in one domain appears in the other as an "innovation." The pleasure of having an insight that does nothing more than deepen understanding of a thing lies completely outside the expert's ken; he aims for practical applications, and in this he is not paralyzed by doubts.

He has a thriving connection to economics. Here also he understands something of the matter; but even more important is his confidence in its language. The language of economics is known to be impenetrable; the business section of the newspaper makes sense only to initiates. The widely accepted fairy tale that economics is as complex as its language seems to indicate is one of his great allies.

Most of the time expertise still locates itself within the tradition of unbroken confidence in the marvels of technique. It appeals to faith in progress or to the fear of a crisis, and it supports this appeal by means of *prognosis.* It employs the method of self-fulfilling prophecy and backs it up with futuristic scenarios: "Freiburg 2000!" "Goals 2000" In this way the experts reveal two paths: one leads, under the flag of "development," up into the heights; the other, which lowers this flag, leads to an inevitable decline, to crisis and catastrophe. The path up is that of *his* economy. He presents commissioned reports under headings like "Perspectives on Future Social Developments"; he says things like "taken as a whole, however, this sector will diminish in importance"; and he knows: "The Future is above all a question of how persuasively the goal is presented, not of the exactitude of biennial predictions" (Späth)—It is generally not the industrialist who speaks in this way. Industry keeps noticeably silent. It gives commissions to public relations agencies, journalists, "Research Institutes on the Configuration of the Future." And it lets the actors in the leading political positions do their PR, entertaining the public with what they are allowed to present as "politics" within the boundaries permitted by industry.

The expert sets universal standards. He does this first of all by attribution: information and communication, development and sexuality appear as properties which *belong* to humans in two ways: naturally, and as a right that can be demanded. They are basic needs and economic programs. The expansion of these amorphous plastic words silently prepares the way for the expert. He strengthens their effect by stigmatizing those who do not correspond to them. He defines the universal standards by drawing a dualistic opposition between "progressives" and "reactionaries," "normal" and "abnormal." He projects an order that turns a majority into useless deviants. They are declassed and stigmatized as "illiterate" and "underdeveloped" or, more recently, "developing." They live below the "poverty line" or without the "minimum compatible with existence." They are backward—insufficiently serviced and insufficiently informed. The world is a set of stairs. His vantage is from the egocentric perspective of those who stand on the top step. From there he projects the rising stairs and ranked orders of his hierarchies into the language. In this way language itself comes to express needs.

Linnaeus designed a special language that made it possible to arrange the riches of nature in a way that could be seen all at once. Among other things he worked out a word field with no gaps, which could precisely distinguish the forms of plants from their roots to their crowns. He laid the groundwork for diagnosing plants. The experts, too, elaborate word fields with no gaps. They produce catalogs of criteria—with the intention of singling out deviants and identifying deficiencies that will require their services.

Germany leads the way in this. Hardly is a new person born before he or she is subjected to tests set out in an "examination booklet for children." The point is the "early recognition" of illness. Underweight? Overweight? Irregular breathing? Macrocephaly? Is the head too large or too small? In the first hours and days of the child's life on earth, sixty-six criteria are checked out. If everything is all right, then on the second page ("E2: Basic Examination of the Newborn") an "x" will be inscribed in the little box marked "no distinguishing characteristics." The booklet is in use for the first four years. Children are compared to developmental formulas, E2 to E8, at prescribed intervals. It is a health passport. A consumer has been born, who must now be checked out and certified. Upon entry into kindergarten and school, apprenticeship, university, and the army, he or she must pass ever more refined tests. Thus does the scarcity of jobs engender a bureaucratic rococo.

Economics and administration are inseparable. Everywhere the expert penetrates, the language of administration expands. Health is a particularly active field of growth. "The increase in health services does not correspond in the slightest with a reduction in death and illness," the *Frankfurter Allgemeine* observes.

What is the effect of this transgression of boundaries and confusion of spheres? Of what use is it, when the crude images of administration wander over into the domains of health and child-rearing, environment and development? This mainly happens in the name of service, and it is a service, but at the same time it also conceals an interest. The administration of service develops a clientele, but at the same time renders it unrecognizable as such. The accent is shifted from the interest of clients to its own "organization." And through this shift, both clients and service providers appear in a new light. Administrative language lends social status to healing aides, educational aides, environmental aides, and developmental aides. The language professionalizes and institutionalizes. More and more transparent expressions can be created: "basic health services," "self-help projects," "community health workers."

Adorno was preoccupied with this word type: to him it had something exhausting in it. "Community health worker" is no longer a word, even less a name; it is a function. Health and environment, developmental aid and science immediately become serious when they appear on the defining screen of administrative language. This language changes intentions into projects and makes projects into institutions. Once these exist—first of all in words—they gain a life of their own. They become commonsensical and attain the inertia of an institution. The name of the establishment and the title do their job. The experts' own language ensures that they won't become superfluous. It keeps the business going and supports an independent agent. Countless careers exist primarily because of their institutionalization in language. Someone formulates a universal law: "Awareness of our universal interdependency is the basis of human responsibility"; describes unimproved life as a pathology: "Insufficient consciousness and insufficient willingness to act are signs of personal and social disease"; requires the virtually impossible: "Be your own boss"; and already an institute has been founded. It only needs a name, for example W.I.L.L. ( = Workshop Institute for Living-Learning, New York) or TCI (System of Theme-Centered Interaction). Language not only makes the man, it also creates the institutions.

If the function of the expert consists above all in the transmission of scientific and technical accomplishments into the practical world, it is easy to understand why he prefers abstractions: they are so encompassing and general that they are easy to transmit. Abstractions are bridges that can link the most incongruous domains. In this they do the work of metaphors and are perhaps, to use a phrase of Nietzsche's, the "residue of metaphor." But they have overcome the drawback of metaphors by sloughing off any pictorial content that might serve as a reminder of their origins. This is why they are so well suited for colonization: they can be transmitted from one domain to another without a hiatus and without drawing attention to any intervening chasm.

Abstract language allows the world to be planned, levels it out evenly, and makes it available to the drawing board. It constructs homogeneous and easily visualized spaces. It avoids sensuousness, diversity, and individual variation, and focuses on what remains when one gets rid of all particular cases. This is precisely how it opens up the world for exploitation.

At the same time abstract language serves to cover up reality. It prevents the imagination from reflecting on what actually happens to people. It ignores what they experience and what they feel, their life histories. The language of the overview leads to disregard of what is most important. The seal of science or of administration, stamped on the everyday by the expert, hides suffering beneath an inhuman objectivity. The expert robs the senses of their reality. In her book about development aid, *Deadly Help*, Brigitte Erler writes:

> But the most important instrument besides selective recognition is abstraction. With its help we can continue to justify our development aid in spite of accumulating knowledge about its harmful effects, and in this way make our atrocities psychically bearable. Abstraction frees us from having to think consequences through to their ends and from having to answer the question: for whom? Even if every day we hear how small farmers are losing their land through our activities, we still maintain that our projects raise the national production levels. I have seen it again and again, how experts dodge to a higher level of abstraction when they notice that the logic of a conver-

sation is leading them to see that because of our help the farmers are starving to death.

Experts often transmit a sense of space—a sense of extensive space. Words such as "global" and "world-encompassing" suit them; so do "the ordering of space," "concentration of industry," and "high-density development." Space itself is developed. The experts define it. They designate a mountain as "suitable for development," ascribe "leisure value" to the "adjacent resort region" of a city, or describe a whole continent as a "developmental region." The only space this kind of expert can recognize is one he can penetrate. So he expands his standardized vocabulary, defines, orders, geometricizes, and leaves behind an altered world.

Compressed air expands when the container of air is opened; at the same time its pressure and the temperature fall. The language of the experts is expansive in this sense; warmth and pressure are lost. Their language thins and empties itself in its expansion.

Experts project large vistas in both time and space. The idea of multiple applications is another example of this. Nothing can be only itself; it must be copiable. A doctor establishes a hospital, a pedagogue a school, a natural scientist undertakes a trial—the expert in question always aims at imitation and expansion. He establishes a model school, a model hospital, or a model project. He does nothing that refers only to itself; and if we just look at the language in which he clothes his acts, we can see in what varied ways he is a strategist.

At a symposium on the development of Freiburg I heard: "These three projects must be tackled." "We are permanently at the cutting edge." Someone spoke about "deploying the leading technologies of the twenty-first century in the upper Rhine basin." The sound as well as the word choice was militaristic. The upper Rhine basin became a mobilization area. The expert is a soldier. Military images openly or covertly inform his language.

Let us now review the profile of the experts which I have drawn with the help of my texts:

—his language has a strategic cast, which is conveyed not only through the feeling it transmits of limited time and expanding space, but also through the idea of multiple applications and through the fact that he declares the whole world his marshaling area;

— the high level of abstraction in his language smooths out the world and simultaneously diverts his gaze from the suffering he has caused. He can also lean on the abstractions of science and administration;

— his language evokes science and administration even when it tries to hide discontinuities between its origins and application. He carries the word fields of science over into the process of administrations. In his hands science becomes a set of administrative acts;

— he also uses science as a resource, an arsenal that supplies the means of altering the everyday. He borrows the prestige that belongs to scientific language; it is useful to him as a means of advertising;

— the expert distills his elixir from a compound of science, economics, and administration. The economic ingredient is often least noticeable; but it covertly does a great deal of the dirty work.

There are two further details that I want to add to my profile:

The expert is an enemy of extremes. He identifies extreme positions as dangerous, and one of his most common arguments is that one must maintain a middle ground between enthusiasm for progress and fear of the future, between economics and ecology. There is such a thing as a car fetish, he says, but there is also a danger of turning the car into a bogeyman: one needs to find a balance between these views. The pattern he follows is a parody of Platonic dialectic. He holds the middle ground regardless of the question and bravely faces into the future: "We must endure the tensions."

I have already said that specialization is the connective tissue of all expert language, no matter what its origin. If a river in a small town floods its banks, a language previously unknown in this region will begin to shoot up like weeds in response to this "catastrophe." "Sewage experts" will be cited. There will be talk of putting people "on alert" and establishing a "control center." The governor of the region will demand that "a permanent solution" be found for the future.

The language of the experts soothes. Specialized expressions have something enormously reassuring about them. They dull the edge. They say: Don't worry, the matter is safely in the hands of those who know. In the case of the "peaceful uses of atomic energy" all that remains is a "residual risk." The expert has the possibility of permanent destruction of districts or countries under control, through his subtle distinctions between a reactor problem and a worst case scenario. He masters the dangers of radiation by invoking the term "safe levels." The language substitutes for the safety agencies, which we know to be no longer effective. Often it serves as an exorcism. There is an accident somewhere: experts immediately arrive on the scene and quote numbers. The expert is a wizard whose language acts as a charm: nothing could be clearer. A soothing facade of language shields us from every tumult. In bad times the ritualized formulas of the news are produced in this way: they ban the dreadful by means of the language of specialized administration.

But here lies perhaps the deepest reason for the proliferation of specialized language observable everywhere. Our basic impulse is to be in motion; the experts contribute in various ways to a general mobilization. But in a world so thoroughly mobilized one also needs some people who provide security and perform exorcisms. The language of the experts imitates stability and secures the journey into the future.

In the previous chapter we spoke briefly about the fact that the language of experts seems to be influenced in many ways by plastic words, and observed that these building blocks exert a gravitational pull on any word group in which they are placed. Proving this would involve showing how the expert's vocabulary has changed in the environment of plastic words. It would require a semantic analysis of sentences, which showed that the synonyms of a given plastic word had been displaced. The same point can be made in an abbreviated way by demonstrating that the criteria we have established for plastic words apply to the expert as well. Let us see whether the expert can replace the plastic words in our composite image.

*The expert*
— silences
— and reforms the world of everyday life through the concepts and the vocabulary of the scientific world;
— he eliminates the chasm between these spheres (1–5);

— his language has a very wide radius of application
— and displaces locally meaningful signs (6–9);

—his speech is poor in content;

—his speech reduces great diversity to a common denominator (10–12);

—he disembeds localities from history,

—transforms them into a laboratory,

—and dispenses with the question "good" or "bad" in favor of the question "progressive" or "backward" (13–15);

—he mediates goods and always appears on the side of Enlightenment;

—the resonance of the name "expert" and the social function he fulfills are more important than what he actually does (16–18);

—he awakens limitless needs,

—whose "naturalness" become an imperative through him;

—he is capable of replicating things (otherwise he is replaced) (19–22);

—the efforts of expert bodies raise prestige;

—he institutionalizes himself and the need for his help through his language;

—he creates compound words and new words which serve as flexible instruments with which to manufacture new models of reality (23–26);

—he makes the past look out-of-date,

—his position is relatively new,

—he has the cachet of the international (27–29),

—and his language lacks an individual voice (30).

It might be felt that by this substitution I have only found the Easter egg I hid myself: that I sketched the profile of the type who serves the process of civilization on the basis of his language, observed that the foundation of this language is a special group of "plastic words," and then concluded by comparison that the expert's language is very similar to those words. Granted: the conclusion is a tautology. But this tautology is a verifiable reality.

The conclusion would be even more compelling if it could be shown that the concept of mathematization accurately describes what plastic words and experts have in common. I will attempt this demonstration in my final chapter.

# 5

## The Mathematization
## of the Vernacular

What has been said up to now can be summarized in one sentence: our vernacular is becoming increasingly mathematized. By this I mean a kind of simplification that I want to describe still more precisely. A further proposition: once the vernacular has been mathematized, it is predisposed to accept the computer. The computer is the consequence and the expression of this process and its complement.

In 1931 Rudolf Carnap employed the term *Sphärenvermengung* ("mingling of spheres") in an essay directed against Martin Heidegger's lecture "What Is Metaphysics?" He was referring to the linking together in one proposition of two things from separate and incompatible logical spheres. In his famous example he declared that "Caesar is a prime number." Such a sentence is said to be not incorrect, merely devoid of meaning. It fuses two incompatible elements.

Carnap says that Heidegger is a particularly egregious example of a mingler of spheres, a writer of sham sentences. Carnap regarded him as an artist who had stumbled onto philosophy, a closet musician who had chosen the wrong medium. "Metaphysicians are musicians without musical talent." This essay was written before Heidegger turned to poetry and shows considerable prescience.

As he goes on, though, Carnap proposes another kind of mingling of spheres, though of course, without calling it that. He claims universality for the language of physics. "Every statement of science can in principle be shown to be a statement of physics. . . . Because the language of physics is becoming the basic language of science, the whole of science is becoming physics."

This means, in other words, that the whole of science is becoming mathematical—an old idea already proposed by Bacon and Kant. Carnap explicitly includes not only biology but also psychology, sociology, history, culture, and economics: "All disciplines belong to the unified science of physics."

This idea might appear a bit nutty if we did not know that it has rapidly gained ground and will continue to do so. Even the humanities lie within its scope, though how far they have already been affected is not a subject I can address here.

Our theme is a different and more thoroughgoing mingling of spheres: the sculpting of the everyday world and its language by the natural sciences. This has become a clearly observable phenomenon, particularly in the last two decades. The gulf separating the everyday world from the world of science, whose purest expression in our age is mathematics, is jumped over or "passed over" as though it hardly existed. What appears in one world as an extension of knowledge or as a technical achievement surfaces in the same guise in the other world, as if the everyday world had no autonomy and no quite different criteria of its own. This mingling is *the* problem.

The leap from the sciences into the everyday world is made easier by what looks like a common language bridging the chasm between the two spheres. The broadest "bridges" are the plastic words.

Despite the fuzzy status of this relatively young class of words, they appear to be (to use the risky expression) *mathematizing* our everyday language. This should not be understood too literally. I am using the word "mathematizing" tentatively to indicate the direction in which science seems to be leading the vernacular, the common features in this movement and its logical outcome. In naming the characteristics of plastic words, I use terms that can also be employed in describing the language of mathematics. To prove what I am saying, I will put my finger on six shared characteristics and briefly summarize them:

1. Plastic words are characterized by a high degree of abstraction. This abstract language creates homogeneous domains that can be scanned quickly. It directs attention away from individual differences. "The Federal Republic is becoming an information society." This language delivers the world into the hands of planners, levels the terrain, and places everyone at the mercy of the drawing board.

Mathematics is the most abstract form of art.

2. Words such as "communication" lack a historical dimension; they are not embedded in any particular place or society. They are shallow and they taste of nothing. They describe nature in the terms of natural science. They banish history from the worlds they invade and destroy the human scale.

For the sake of contrast we could think of examples in which language is closely tied to its context: the Latin of the Middle Ages, the language of an oral culture, the German *Bildungssprache* of around 1800. In these cases words were used within definite horizons, but at the same time had the capacity to convey experience or insight on a human scale.

Amorphous plastic words release language from such ties. They evoke no particular setting; they are universal. They recast life histories as natural processes and say that everything is basically the same — another module.

When history is regarded with the eye of a physicist, as a domain always and everywhere the same, its effects on us become as inexorable as nature's. It is as if history were actually driven to take on the character ascribed to it.

Mathematization is the most radical appellation for such a procedure. Mathematics is *the* ahistoric, universal form of art, independent of time and place.

3. Our key words are used in the manner of boldly outlined building blocks, as though one were dealing with numeric quantities. The aura

predominates, but even in the vernacular, many of these stereotypes suggest a quantifiable amount. Not just "energy," "production," and "consumption," but even "information" or "communication" increasingly appear before us as statistics.

4. The plastic words have a tendency to create sentences when placed in almost any order. The words are alarmingly interchangeable, they can be equated with one another or strung together in a chain of equations. "Communication is exchange. Exchange is a relationship. A relationship is a process...."

The motility of these words, their capacity to combine with other terms, seems almost unlimited; the possibilities of using them seem infinite.

The linguistic principle of arbitrariness, seen in the currently popular idiom "I will start from the position that," i.e. "I will assume that" [ich gehe davon aus], the capacity to switch positions and perspectives in an instant, is presumably much more common in contemporary usage than in that of earlier ages. Our language is quite fragile, has little native resilience, and seems to be almost without resistance to the new compounds churned out every day: "information society," "a guaranteed future for the media" [gesicherte Medienzukunft]. Where power or material interests are at stake, for example in corporations with a turnover of tens or hundreds of billions, any kind of nonsense will serve. In science, words are made arbitrarily precise; in the vernacular the same words are merely arbitrary.

5. We are speaking about only a small set of words, but they are the building blocks of countless models of reality. Whether the topic is the Third World, health, agriculture, or town planning—in the mill of the plastic words models are manufactured and projects developed in an instant. Experts inflect each word according to the sector to which it is assigned. Some of the words are already on the way to becoming suffixes, to entering a grammatical category. They tend to form series. Our world is deficient, malleable, and continually refashioned into new structures: this is the point of this modular Lego language.

Producing a variety of models by combining and multiplying a limited set of symbols has something arbitrary and something economical about it. It is also somewhat reminiscent of simple mathematics.

6. Mats of words creep across the surfaces of our living spaces and grip them fast. They are comprised of the advice of experts, catalogs of criteria, examination papers, marks, points, tests, test results, and percentages. Enumeration and geometrization reach into every crack. "The hierarchy of substances is destroyed, a single one replaces all the others: the entire world can be plasticized" (Roland Barthes).

Abstraction, ahistorical universality, numerical sizing, the arbitrariness principle, reduction, combination, multiplication, model building unrestricted by any social norm, geometrization, and enumeration—if these expressions fit, then it is meaningful to speak of mathematization of the vernacular. Mathematics's claim to universality has not only reached the humanities, but has also jumped over into everyday society, where it is mirrored in its speech. But the vernacular has not become more precise as a result: plastic words only parody mathematics, just as they only denature the vernacular. Carnap's example of the mingling of spheres no longer seems accurate: Caesar *is* a prime number.

The phenomenon has two striking parallels: I see the first parallel in the new concept of language put forward by several leading twentieth-century linguists, which I have already sketched briefly. This concept reflects and supports the mathematization and mobilization of the vernacular. It began in 1906 with de Saussure's statement of the principle of the arbitrariness of the linguistic sign—an ahistorical notion like that of the social contract. It continued with the publication of Charles K. Ogden's proposal in 1930 for what he called Basic English—he proposed a model language comprising 850 English words, a *lingua franca* whose small all-purpose lexicon was to be diversified by combination. And it completed itself with Chomsky's mathematical syntax theory of 1957 (1965), in which the capacity to generate and understand an infinite number of sentences, using a limited group of symbols and a small set of rules for combining them, became the guiding principle of a grammar. This principle is combined with the idea that all languages basically follow the same pattern. The principle of free combination limited only by the requirement of "semantic congruence" connects with a principle of ahistoric universality. Chomsky's theory of syntax is the most advanced formulation of a mathematical concept of language and thus the most adequate expression of our age. That its worldwide success coincided with the most intensive period of world colonization ever seen may have been more than a "coincidence."

The second parallel is with Orwell's "Newspeak." Orwell was fascinated by Ogden's Basic English for a short time, and he worked a parody of this concept into his dystopia Nineteen Eighty-Four. What he describes in the novel and in the appendix as the "short grammar" of Newspeak is a language without a historical dimension, an artificially created planlanguage sanctioned by the state. Its more general characteristics are as follows: reduction of the vocabulary and standardization of meaning, the manipulation of boldly outlined building blocks (words which still evoke earlier history are eradicated or pruned); economy, made possible by the free combination and transformation of the elements (doubleplus-cold = very cold, doubleplusuncold = very warm); a binary system of values (old = bad, new = good); arbitrariness of meaning (war is peace, peace is war); logical transparency and uniform geometrical norms (he thinked); abstraction and abbreviation. In short, Newspeak is a mathematized form of colloquial language, a particularly cheerless form: the language of a totalitarian state completely divorced from history.

Our theme is a different instance of mathematization: the basic international code of modular plastic words. It is simple, has no real history, is easy to learn and manipulate, and is limited in vocabulary and syntactic rules: a type of Lego. It overlays and displaces the local vernacular, replaces nuanced and nonverbal modes of expression, and over time insinuates itself everywhere. The Lego language of the industrial state plasticizes the planet.

Mathematics, within its own domain, is a pure and universal art. Almost like music, it appears to be an absolute language. At least it is far removed from the historicity of the everyday world. When transferred into the vernacular, its purity becomes a void. This is not meant to detract from mathematics, but only to point out how successfully it has been transplanted. The principle of universality begins to erode the terrain of the vernacular long before it has been completely leveled. Scientific terms like "communication," "information," and "sexuality" are the harbingers in the everyday world of a growing desert.

Why is it that, in modern times, whatever exists as substance, matter, or material invites transformation? What is it that makes the stuff of our world purely functional, not just occasionally or temporarily but on principle? What is it that turns such diluted concepts as "exchange," "role," and "decision" into basic tools of practical life? They are abstract and empty and appear hardly human, or rather they constantly over-shoot the human sphere. They annul ethics in the everyday world. A metaphorical colonization distorts life and transforms it into a laboratory. The laboratory works in infinitum. Communication, information, sexuality: these are after all idealizations, potentially infinite in their abstractness, which are ascribed to human beings. Mathematically idealized nature, as attributed to us, becomes an endless program all on its own. Corresponding to this is the peculiar motility in our speech, its unbalanced quality, and its constant tendency to petrify into enveloping formulas.

At this point speculations about the effect of the computer become relevant. The computer is a tool of total mobilization. Its methods seem precise but are wildly out of control. It is the most versatile calculator conceivable, the long arm of mathematized society. The vernacular is predisposed to receive it and after its universal adoption, which is already well under way, the vernacular will presumably continue to change in the same way.

Our theme is touched on by Theodor Adorno and Max Horkheimer in *Dialectic of Enlightenment*: "By its anticipatory identification of a fully elaborated mathematization of the world with truth, the Enlightenment believed itself safe from the return of the mythic. It treated thought and mathematics as one, thereby unleashing the latter and endowing it with absolute authority. . . . The mathematical approach became, as it were, the ritual of thought."

Even more important, if I read it correctly, is a lecture that Edmund Husserl gave in Vienna in 1935 and later elaborated into a fragmentary work, published posthumously, called "The Crisis of European Society and Transcendental Phenomenology."

"O man full of arts," says the Egyptian king to Theuth, the divine inventor of number and arithmetic, measurement and astrology, board and dice games and the alphabet, after Theuth shows him these arts, desiring them to be passed on to all Egyptians. "O man full of arts, to one it is given to create the things of art, and to another to judge what measure of harm and of profit they have for those that shall employ them" (*Phaedrus*, 274). We are in need of this other man. Politics lacks independence, and so does the vernacular realm of simple intransitive living. We are the victims of a totalitarian monism of the natural sciences and their technological elaborations [Transformatoren]. Language mirrors and allows this domination.

In the ancient Egyptian fable that Socrates recounts in the *Phaedrus*, invention and judgment are completely separate. Today they pass seamlessly into each other: a common language covers up the fundamental differences between communication in science and communication in everyday life. Our language is laid waste and its boundaries violated. It is robbed of the layers of experience through which it expresses what is human.

In 1933, at the Chicago World's Fair, a slogan of devastating simplicity was coined: "Science discovers, technology exploits, man adapts." That is how it is. Regrettably.

# Appendix
## Characteristics of Plastic Words

## A. Origin and Usage

1. The speaker lacks the power of definition; the words do not acquire meaning or nuance from their contexts.

2. As "context-autonomous" words that do not depend on their connections, they superficially resemble the terms of science, but lack the precisely defined meanings of such terms, and their freedom from associations. The use of the same word inside and outside science leads to the assumption of kinship, and to the words becoming independent norms. In the vernacular, these nephews of science become stereotypes.

3. As a rule they originate in the vernacular, are adopted and reshaped by some brand of science, and then, like returning émigrés, rejoin the vernacular.

4. They have the character of metaphors inasmuch as they link the heterogeneous spheres of science and everyday life. They are distinct from metaphors in that they no longer evoke any image; they do not, like other comparisons, indicate their origin.

5. This makes their capacity to alter and illuminate their objects even more powerful. The less obvious their metaphorical character, the less it is noticed, and the more effectively it works. These words become commonsense, background concepts in our thinking.

# B. Scope

6. The words surface in countless contexts. Their application is limited hardly at all by space or time.

7. They squeeze out and replace a wealth of synonyms. Synonyms after all are not words whose meaning is the same but whose meaning is similar, words with as many delicate differences and shadings as there are contexts. Before plastic words one knew which synonym belonged in which factual or social context. Now there is a "jack of all trades," a word that serves the whole world.

8. They squeeze out and replace the *verbum proprium*, which precisely "fits" in a given context, with a nonspecific word.

9. They fill silences and replace indirect ways of speaking, exposing delicacy and tact to the action of stereotyped generalities.

# C. Content

10. When we seek to grasp the meaning of the words, through their content rather than their sphere of influence, it comes down to a single characteristic. They manifest the logical law of the inverse proportionality of extension and intention: the broader the application, the smaller the content; the poorer the content, the larger the application. They are words that reduce a gigantic area to a common denominator. They put forward a universal claim, with a reduced and impoverished content.

11. In other words, the object spoken about, the referent, is not easy to grasp; the words are poor in substance, if not altogether without substance.

12. They seem to resemble the concepts of postclassical physics: purely imaginary, meaningless, self-referential, and functioning only as stackable poker chips. Is language being undermined in parallel with the use of these poker chips in the thought structures of mathematics and physics?

# D. History as Nature

13. The words lack a historical dimension; they are embedded in no particular time or place. In that sense they are shallow; they are new and they don't taste of anything.

14. They reinterpret history as nature and transform it into a laboratory.

15. They dispense with questions of good and evil and cause them to disappear.

# E. Power of Connotation and Function

16. Connotation dominates, spreading out in expanding waves. In place of the power of denotation, they provide an experience of counterfeit enlightenment.

17. Their connotation is positive; they formulate a property or deliver the illusion of an insight.

18. In their usage the *function* of the discourse dominates, not its *content*. These words are more like an instrument of subjugation than like a tool of freedom.

# F. General Function

19. By means of their limitless generality they give the impression of filling a gap and of satisfying a need that had not previously existed. In other words, they awaken a need. They reduce all domains to a common denominator and sound an imperative and futuristic note. The words seem to demand that these domains adjust themselves to the words and not vice versa. They draw attention to deficits.

20. Their asocial and ahistorical naturalness reinforces this demand.

21. Their powerful aura of associations demands action.

22. Their many-sided generality brings about consensus.

# G. Social and Economic Usefulness

23. Their use distinguishes the speaker from the unremarkable world of the everyday and raises his social prestige; they serve him as rungs on the social ladder.

24. They carry the authority of science into the vernacular: they enforce silence. (In the GDR Marxist-Leninist science was already monumentalized by being the explicit foundation of the state structure. In the Federal Republic the scientific vocabulary pushed itself into a comparable position as an instrument for awakening economic needs.)

25. These words form a bridge to the world of experts. Their content is actually no more than a white spot, but they transmit the "aura" of another world, in which one can obtain information about them. They anchor, in the vernacular, the need for experts. They are pregnant with money. They command resources, and, in the hands of experts, become resources.

26. They can be freely combined, and they are eager to increase themselves through derivation and the creation of compounds. This modular

capacity makes them an ideal instrument in the hands of experts interested in the speedy manufacture of models of reality.

# H. Time and Place of Dissemination

27. Their scientifically authorized objectivity and universality make the older words of the vernacular appear ideological. A word like "communication" makes alternatives—conversation, discussion, gossip—suddenly appear out of date.

28. The words appear as a new type. In recent history such newcomers have evidently been introduced in each epoch. The type in vogue in the 1930s is not the type in vogue in the 1990s.

29. This vocabulary, even if it appears at slightly different times in different places, is international.

# I. Connection to Making Oneself Understood without Words

30. The words cannot be made clearer by tone of voice, pantomime, or gesture, and cannot be replaced by these.

# Bibliography

This bibliography is a selection from the literature on which I have directly or indirectly drawn. For this edition, English translations have been substituted when available.

Adorno, Theodor W. 1974. *The Jargon of Authenticity*. Translated by Knut Tarnowski and Frederic Will. Evanston, Ill.
———. 1974. "Wörter aus der Fremde." In *Noten zur Literatur*, vol. 11 of *Gesammelte Schriften*. Frankfurt am Main.
Adorno, Theodor W., and Max Horkheimer. 1972. *Dialectic of Enlightenment*. Translated by John Cumming. New York.
Anders, Günther. 1956. "Die Welt als Phantom und Matrize. Philosophische Betrachtungen über Rundfunk und Fernsehen" (The world as phantom and matrix. Philosophical observations about radio and television). In *Die Antiquiertheit des Menschen*. 2 vols. Munich, 1980.
Arendt, Hannah. 1977. "Die Lüge in der Politik." *Neue Rundschau*. Jg. 1977, p. 190.
Barthes, Roland. 1972. *Mythologies*. Selected and translated by Annette Lavers. New York.
Behn, Hans Ulrich. 1971. *Die Regierungserklärungen der Bundesrepublik Deutschland*. Munich.
Beyme, Klaus von. 1979. *Die großen Regierungserklärungen von Adenauer bis H. Schmidt*. Munich and Vienna.
Brockhaus-Wahrig. 1980–1984. *Deutsches Worterbuch in sechs Bänden*. Hersg. Gerhard Wakrigt, Hildegard Krämer, Harald Zimmermann. Stuttgart.
Bühler, Karl. 1990. *Theory of Language: The Representational Function of Language*. Translated by Donald Fraser Goodwin. Philadelphia.
Carnap, Rudolf. 1931. "Überwindung der Metaphysik durch logische Analyse der Sprache." *Erkenntnis* 2: 219–41.
Chomsky, Noam. 1965. *Aspects of the Theory of Syntax*. Cambridge, Mass.
Coseriu, Eugenio. 1970. "System, Norm, Rede." In *Sprache und Strukturen und Funktionen: XII Aufsätze*. Tübingen.
Coulmas, Florian. 1985. *Sprache und Staat. Studien zur Sprachplanung*. Berlin and New York.
Dittmann, Jürgen. 1984. "Sprachlenkung und Denkverbot. George Orwell als Sprachkritiker." *Freiburger Universitätsblätter* 23: 31–47.
Duden. 1976–1981. *Das grosse Wörterbuch der deutschen Sprache*. Hersg. Günther Drosdowski Mannheim. 2 Auflage in acht Bänden.
Eppler, Erhard. 1992. *Kavalleriepferde beim Hornsignal. Die Krise der Politik im Spiegel der Sprache*. Frankfurt am Main.

Erler, Brigitte. 1985. *Tödliche Hilfe. Bericht von meiner letzten Dienstreise in Sachen Entwicklungshilfe.* Freiburg im Breisgau.

Esteva, Gustavo. 1985. "Development: Metaphor, Myth, Threat." Unpublished manuscript.

Foucault, Michel. 1971. *The Order of Things: An Archaeology of the Human Sciences.* New York.

Freud, Sigmund. 1921. *Group Psychology and the Analysis of the Ego.* In *The Standard Edition of the Complete Psychological Works.* Translated and edited by James Strachey. London.

Garza Cuaron, Beatriz. 1991. *Connotation and Meaning.* Translated by Charlotte Broad. New York.

Gauger, Hans Martin. 1972. "Die Wörter und ihr Kontext." *Neue Rundschau* 83: 432–50.

———. 1972. *Zum Problem der Synonyme.* Tübinger Beiträge zur Linguistik 9. Tübingen.

*Geschichtliche Grundbegriffe. Historisches Lexikon zur Politisch-sozialen Sprache in Deutschland.* 5 vols. (Basic historical concepts). Edited by Otto Brunner, Werner Conze, and Reinhardt Koselleck. Stuttgart, 1972.

Giesecke, Michael. 1979. "Schriftsprache als Entwicklungsfaktor in Sprachund Begriffsgeschichte." In *Historische Semantik und Begriffsgeschichte,* ed. Reinhardt Koselleck. Stuttgart.

Gipper, Helmut. 1962. *Bausteine zur Sprachinhaltsforschung.* Düsseldorf.

———. 1964. "Zur Problematik der Fachsprachen. Ein Beitrag aus sprachwissenschaftlicher Sicht." In *Festschrift für Hugo Moser zum 60 Geburtstag.* Düsseldorf.

Grimes, Barbara F., ed. 1978. *Ethnologue.* 9th ed. Huntington Beach, Calif.

Habermas, Jürgen. 1978. "Umgangssprache, Wissenschaftssprache, Bildungssprache." *Merkur* 32: 327–42.

Hannappel, Hans, and Hartmut Melenk. 1979. *Alltagssprache. Semantische Grundbegriffe und Analysebeispiele.* Munich.

Hassenstein, Bernhard. 1966. "Was ist 'Information'?" (What is information?) *Naturwissenschaft und Medizin* 3: 38–52.

———. 1979. "Wie viele Körner ergeben einen Haufen? Bemerkungen zu einem uralten und zugleich aktuellen Verständigungsproblem" (How many seeds make a pile? Remarks on an ancient yet still current problem in understanding). In *Schriften der Carl Friedrich von Siemens Stiftung,* ed. by Anton Peisl and Armin Mohler. Vol. 1, *Der Mensch und seine Sprache.* Frankfurt am Main.

Heidegger, Martin. 1950. "Die Zeit des Weltbildes." In *Holzwege.* Frankfurt am Main.

Heringer, Hans Jürgen, ed. 1982. *Holzfeuer im hölzernen Ofen. Aufsätze zur politischen Sprachkritik.* Tübingen.

Husserl, Edmund. 1962. *The Crisis of European Sciences and Transcendental Phenomenology: An Introduction to Phenomenological Philosophy.* Translated by David Carr. Evanston, Ill.

Illich, Ivan. 1976. *Medical Nemesis: The Expropriation of Health.* New York.

——. 1984. *Schule ins Museums: Phaidros und die Folgen.* With an introduction by Ruth Kriss-Rettenbeck and Ludolf Kuchenbuch. Schriftenreihe Zum Bayerischen Schulmuseum Ichenhausen, Zweigmuseum des Bayerischen Nationalmuseums, vol. 3.1. Bad Heilbrunn.

——. 1985. *H₂O and the Waters of Forgetfulness: Reflections on the Historicity of "Stuff."* Berkeley, Calif.

Illich, Ivan, et al. 1978. *Disabling Professions.* New York.

Jochmann, Carl Gustav. 1983. *Politische Sprachkritik. Aphorismen und Glossen.* Edited by Uwe Poerksen; selected and annotated by Uwe Poerksen, Siegfried Hennrich, Herbert Klausmann, Eva Lange, and Jürgen Schiewe. Stuttgart.

Jünger, Ernst. 1930. "Die Totale Mobilmachung." In *Sämtliche Werke* II, vol. 7, *Essays* I. Stuttgart, 1980.

——. 1932. *Der Arbeiter. Herrschaft und Gestalt.* Hamburg. Reprint: Stuttgart, 1982.

——. 1934. "Über den Schmerz." In *Sämtliche Werke* II, vol. 7, *Essays* I. Stuttgart, 1980.

Kainz, Friedrich. 1972. "Über die Sprachverführung des Denkens." *Erfahrung und Denken,* vol. 38. Berlin.

Korn, Karl. 1959. *Sprache in der verwalteten Welt* (Language of the world of bureaucracy). Freiburg im Breisgau. Reprint. Munich, 1962.

Kraus, Karl. N.d. *Die Sprache.* Vol. 2 of *Werke von Karl Kraus.* Edited by Heinrich Fischer. Munich.

Ladendorf, Otto. 1906. *Historisches Schlagwörterbuch.* Strassburg and Berlin.

Laplanche, Jean, and J.-B. Pontalis. 1973. *The Language of Psycho-analysis.* Translated by Donald Nicholson-Smith. New York.

Leisi, Ernst. 1961. *Der Wortinhalt. Seine Struktur im Deutschen und Englischen.* 4th ed. Heidelberg.

zur Lippe, Rudolf. 1974. *Naturbeherrschung am Menschen.* 2 vols. Frankfurt am Main. Reprint. 1978.

Lyons, John. 1977. *Semantics.* New York.

Marcuse, Herbert. 1964. *One-Dimensional Man: Studies in the Ideology of Advanced Industrial Society.* Boston.

Möser, Justus. 1968. *Patriotische Phantasien* (Patriotic fantasy). In *Sämtliche Werke. Historisch-Kritische Ausgabe,* vol. 6. Oldenburg and Hamburg.

Nietzsche, Friedrich. 1964. *Complete Works.* Edited by Oscar Levy. New York.

Oberließen, Peter. N.d. "Neuigkeiten aus der Akademie von Lagado." Unpublished manuscript.

Ogden, Charles Kay. 1944. *The System of Basic English.* New York.

Ohly, Friedrich. 1960. "Vom Sprichwort im Leben eines Dorfes." In *Volk, Sprache, Dichtung. Festgabe für Kurt Wagner.* Edited by Karl Bischoff and Lutz Röhrich. Gleßen. Reprinted in *Ergebnisse der Sprichwörterforschung,* ed. Wolfgang Meder. Bern and Frankfurt am Main, 1978.

Olschki, Leonardo. 1919, 1922, 1927. *Geschichte der neusprachlichen wissenschaftlichen Literatur.* Vol. 1, *Die Literatur der Technik und der angewandten Wissenschaften vom Mittelalter bis zur Renaissance.* Heidelberg. Vol. 2, *Bildung und*

*Wissenschaft im Zeitalter der Renaissance in Italien.* Leipzig. Vol. 3, *Galilei und seine Zeit.* Halle an der Saale.

Ong, Walter J. 1982. *Orality and Literacy: The Technologizing of the Word.* New York.

Orwell, George. 1948. *1984.* London.

Paul, Hermann. 1970. *Prinzipien der Sprachgeschichte.* Text edition of the 8th (1968) edition. Tübingen.

Plato. 1964. *Sämtliche Werke 2.* Translated by Friedrich Schleiermacher. Reinbek bei Hamburg.

Poerksen, Uwe. 1986a. *Deutsche Naturwissenschaftssprachen. Historische und kritische Studien.* Forum für Fachsprachenforschung 2. Tübingen.

———. 1986b. "Die Reichweite der Bildungssprache und das szientistische Selbstmißverständnis der Sprachwissenschaft." In *Deutsch als Wissenschaftssprache,* ed. Hartwig Kalverkämper and Harald Weinrich. Forum für Fachsprachenforschung 3. Tübingen.

Polenz, Peter von. 1963. *Funktionsverben im heutigen Deutsch Sprache in der rationalisierten Welt.* Beihefte zum Wirkenden Wort 5. Düsseldorf.

———. 1967. "Sprachpurismus und Nationalsozialismus. Die Fremdwortfrage gestern und heute." In *Germanistik-eine deutsche Wissenschaft.* Frankfurt am Main.

———. 1978. *Geschichte der deutschen Sprache.* Berlin.

———. 1985. *Deutsche Satzsemantik. Grundbegriffe des Zwischen-den-Zeilen-Lesens.* Berlin.

Porzig, Walter. 1930. "Die Leistung der Abstrakta in der Sprache." In *Blätter für deutsche Philosophie 4.* Reprint. In *Das Ringen um eine neue deutsche Grammatik,* ed. Hugo Moser. Wege der Forschung 25. Darmstadt, 1969.

Pross, Harry. 1981. *Zwänge. Essay über symbolische Gewalt* (Duress: An essay concerning symbolic force). Berlin.

Saussure, Ferdinand de. 1974. *Course in General Linguistics.* Edited by Charles Bally and Albert Sechehaye in collaboration with Albert Reidlinger. Translated by Wade Baskin. Revised edition. London.

Schiewe, Jürgen. "Der Begriff des 'Vorlaufs' in der Geschichte. Wessen Vorläufer sind die Amöbenwörter?" Unpublished manuscript.

Schlieben-Lange, Brigitte. 1983. *Traditionen des Sprechens. Elemente einer pragmatischen Sprachgeschichtsschreibung.* Stuttgart.

———. 1987. "Das Französische – Sprache der Uniformität." *Zeitschrift für Germanistik* 8: 26–38. Leipzig.

Schmidt, Wilhelm. 1963. *Lexikalische und aktuelle Bedeutung. Ein Beitrag zur Theorie der Wortbedeutung.* Schriften zur Phonetik, Sprachwissenschaft und Kommunikationsforschung 7. Berlin.

Späth, Lothar. 1985. *Wende in die Zukunft. Die Bundesrepublik auf dem Weg in die Informationsgesellschaft* (Turning toward the future: the Bundesrepublik becomes an information society). Hamburg.

Steinhoff, William. 1975. *George Orwell and the Origins of 1984.* Ann Arbor, Mich.

Sternberger, Dolf, Gerhard Storz, and W. E. Süskind. 1957. *Aus dem Wörterbuch des Unmenschen.* Hamburg.

Ullmann, Stephen. 1959. *The Principles of Semantics*. 2d ed. Glasgow.
Weber, Max. 1919. "Science as a Vocation." In *From Max Weber: Essays in Sociology*, ed. H. H. Gerth and C. Wright Mills. New York, 1946.
Weiland, Wolfgang. 1972. "Entwicklung." In *Geschichtliche Grundbegriffe: Historisches Lexikon zur Politisch-sozialen Sprache in Deutschland* (Basic historical concepts), ed. Otto Brunner, Werner Conze, and Reinhardt Koselleck. Stuttgart.
Weinrich, Harald. 1964. "Typen der Gedächtnismetaphorik." In *Archiv für Begriffsgeschichte* 9: 23–26.
———. 1966. *Linguistik der Lüge. Kann Sprache die Gedanken verbergen?* Heidelberg.
Weisgerber, Leo. 1958. *Verschiebungen in der sprachlichen Einschätzung von Menschen und Sachen*. Cologne.
Weizsäcker, Carl Friedrich von. 1959. "Sprache als Information." In *Die Sprache*. Munich.

# Index

Printed in the United Kingdom
by Lightning Source UK Ltd.
124726UK00002B/20/A

9 780271 024929